Simple
Earth Science
Experiments
with Everyday
Materials

● ● ● ● ● ●

Louis V. Loeschnig

Illustrated by Frances Zweifel

Sterling Publishing Co., Inc. New York

Edited by Claire Bazinet

Library of Congress Cataloging-in-Publication Data

Loeschnig, Louis V.
    Simple earth science experiments with everyday materials / by Louis V.
Loeschnig; illustrated by Frances Zweifel.
        p.   cm.
    Includes index.
    Summary: Presents information on such topics as seismology,
botany, environmental sciences, gravity, and the atmosphere,
with various experiments and activities.
    ISBN 0-8069-0898-X
    1 Earth sciences—Experiments—Juvenile literature. [1. Earth
sciences—Experiments. 2. Experiments.] I. Zweifel, Frances W., ill.
II. Title.
QE29.L68   1996
550´.78—dc20   95-53977
                                                                    CIP
                                                                    AC

10 9 8 7 6 5 4 3 2 1

First paperback edition published in 1997 by
Sterling Publishing Company, Inc.
387 Park Avenue South, New York, N.Y. 10016
© 1996 by Louis V. Loeschnig
Distributed in Canada by Sterling Publishing
% Canadian Manda Group, One Atlantic Avenue, Suite 105
Toronto, Ontario, Canada M6K 3E7
Distributed in Great Britain and Europe by Cassell PLC
Wellington House, 125 Strand, London WC2R 0BB, England
Distributed in Australia by Capricorn Link (Australia) Pty Ltd.
P.O. Box 6651, Baulkham Hills, Business Centre, NSW 2153, Australia
Manufactured in the United States of America

Sterling ISBN 0-8069-0898-X Trade
                0-8069-0365-1 Paper

Contents

Before You Begin

Unlike some Earth books, this one will give you a world of information, dozens of earthly exciting activities and experiments and teach you the how's and why's of becoming a real conservationist—a person who is "Earth conscious" and does everything possible to save and protect our lands, forests, and waters.

You'll learn through experimentation how plants give off oxygen and moisture and how, without them, life could not exist. Included are many leafy lessons and experiments. Also, learn how magnetism and electricity are related Earth forces. Make a pin on a thread sway like a dancing cobra and learn how business security people catch fake coins, or slugs, in vending machines—both by using magnetic force.

You'll see how earthquakes are produced and even make a homemade seismograph, an instrument for measuring them, and you'll build a glacier model that melts, moves, and leaves behind the sand and rock it carries.

Learn about soil, sand, the sun, and fossils, and then make a down-to-earth water filter, solar water heater, and a different type of "chemistry volcano" that foams, steams, and hisses.

As a conservationist, you'll learn how to conserve, save and reuse old clothes, toys, and other household materials. We'll show you not only how to recycle cans, bottles and newspapers but how to recycle your garbage through worm composting—a new and friendly way of reusing Mother Earth.

In addition, you'll learn through facts and experimentation about ozone, fossil fuels, acid rain, rain forests, and global warming—and yes! You can even recycle old newspapers to make your own paper, note cards, and a pinhole camera, or see the constellations on a wall using a pinhole light-box projector.

It is suggested that materials you use for your experiments be reused, recyclable, or biodegradable (can break down, rot, and become part of the Earth again). In most cases, with the exception of a few plastics and paper products, most materials used in this book are recyclable.

All the materials needed for the projects in this book are inexpensive and easy to find—and can be found in supermarkets, variety stores, and drugstores.

Not all materials needed for the projects are listed

each time. As some are used for several experiments, it is suggested that you keep, buy, and save the following to have on hand: various-size bottles and jars, coffee cans, shoe boxes, small plastic or clay garden pots, thermometers, coffee filters, medicine droppers, newspaper, paper clips, spray bottles, quart milk cartons, a magnifying hand lens, toy compass, bar and horseshoe magnets, scissors, string, pencils, paper, protractor, potting soil, gravel, clay, and sand.

All experiments have been simplified and thoroughly tested—they do work! However, some, such as those with living plants and seeds and on ecology, are long term—they take time and patience.

It is best to check with an adult before using or taking any materials needed for your experiments. We'll also let you know of any experiments involving construction, and alert you to safety concerns.

This book was designed with our Earth, and your enjoyment, in mind. It provides dozens of exciting, fun activities that we absolutely guarantee you'll enjoy digging!

Happy Earth Experimenting!

WORLDLY

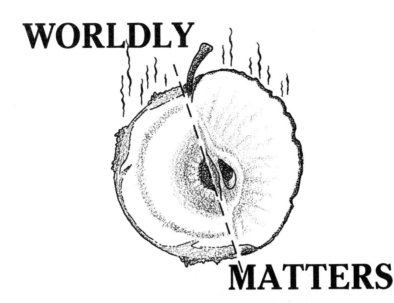

MATTERS

The Earth—a huge ball with an outer crust, inner mantle, and core—travels through space, as do the sun, stars, and other planets. Besides this movement through space, the surfaces of the Earth are also changing constantly. High mountains and deep valleys, both on the land and under the oceans, are all part of the Earth's movement—nothing stays the same. Think of the Earth as an apple sitting in the sun. As the sun warms and dries the apple, its water is lost and the apple shrinks and wrinkles.

While the Earth is not an apple, it undergoes similar changes. The Earth, like the inside of the apple, shrinks or contracts. As the hot interior parts of the Earth cool and shrink, the outside covering is forced to move. The apple's surface makes wrinkly peaks and valleys and, similarly, the Earth's crust forms mountains, valleys, and breaks or cracks called faults.

In this chapter, we'll take a look at some of the forces that affect our Earth, as well as other worldly matters.

Earthquakes: They're Definitely Not Your Fault!

The pressures within the Earth cause great forces, which, in turn, break and crack the Earth's crust. These cracks are called faults, and movement along a fault produces earthquakes. How does this happen?

You need:
3 similar-size hardcover books

What to do:
With the three books held firmly together, bring them close to your chest, book spines (with titles) upward. Reaching under, push upward on the middle book so that it slides upward between the two outer books. Do this several times to make a smooth, straight lift.

Next, firmly hold the books out away from your body, keeping them tightly and evenly together, or aligned. Hold them sideways again, with the titles up and the pages going down. You'll have to apply much force to keep them

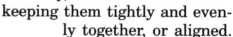

from slipping. Now, release some of the pressure so that the middle book slips.

Finally, hold the books evenly together, spines upward, and rest them on a table. With your hands holding only the two outer books, slide them back and forth.

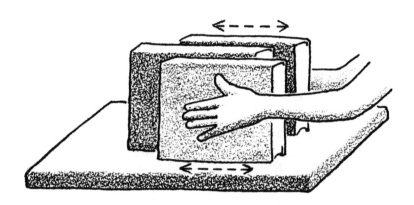

What happens:
The different movements of the books resemble earthquake faults, with much uplifting and slipping.

Why:
In the first two experiments, in which you held the books first close to your chest and then away from your body, you demonstrated dip slip fault movement, a repositioning up and down. The middle book that was forced up (thrust fault) and the one that slipped down (normal fault) are good examples of this type of fault. The books that rested on the table and were moved to slide past one another show the action of a strike slip fault. In this type of fault, movement is sideways (side by side) or parallel.

GET THE LEAD OUT!–BUILD A SEISMOGRAPH SHAKER-MAKER

That's right! With a sharp lead pencil with eraser, you can build a simple seismograph—an instrument used by seismologists (earthquake scientists) to record the strength or intensity of earthquakes. (Adult help may be needed.)

You need:

shoe box with lid
heavy weight for box
pencil with eraser
weights for pencil
  —nails, washers, etc.
clay

masking tape
2 paper clips
string
scissors
2 sheets of paper

What to do:

Carefully cut a tiny slit in the middle near one end of the shoe box lid. Place the open box upright, on one end, and something small and heavy inside to keep it in position. Tape the lid onto the top of the box forming a T. (It doesn't matter if the open part of the box or the bottom of it is towards the slit in the lid.)

Now, place the weight(s) near the tip of the pencil point, but do not cover it, and tape or fasten them on securely. A small piece of clay around the pencil near the taped weights will keep weights from slipping off. The weights must be fairly heavy so the seismograph recorder pencil will make good contact with the paper and draw fairly dark drag lines on it.

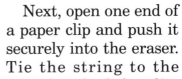

Next, open one end of a paper clip and push it securely into the eraser. Tie the string to the unopened end of the clip. Attach the second paper clip to the other end of the string. Wind the string around the paper clip, much as you would wrap a kite string around a stick. Slip the top clip through the slit and adjust the pencil marker so the tip rests on the table, not perfectly straight but dragging as it moves. Slip the remaining string under one side of the clasp to fasten the upright pencil in place.

Now, cut each paper sheet lengthwise into three strips. These strips will act as roll paper and will record your "earthquake movements."

Place a paper strip against the box (below the slit you made in the lid) and pull the strip forward slowly. Notice how straight the drawn line is as you move the strip of paper.

Next, have a friend bump and shake the table as you pull the paper strips under the dragging pencil marker. Your seismograph makes sideways and up and down movements. Compare the separate strips of paper, how do the lines differ? How do they show the effects of a dip slip fault versus a strike slip fault? (See "Earthquakes: They're Definitely Not Your Fault!")

You Can Move Mountains

Have you ever wondered how mountains are formed? One way is the great pressures from deep within the Earth that cause folds or waves to appear on its surface. These forces cause fold mountains. You can reproduce a version of this force and effect in a simple and easy way. All you'll need is a lump of clay...and a lot of imagination.

You need:

some clay   newspaper

What to do:

Lay some newspaper on a worktable. Place a lump of clay on it and make a clay rope by rolling the lump back and forth over the table with your fingers. When the rope is about 8 inches (20 cm) long, lay it out flat on the newspaper and try pushing inward on the ends, trying to make hills and valleys.

When you're done doing that, roll and smooth out the rope again and place different forces on it. Try to make it bend in new and different ways.

What happens:

The clay rope, with its outside forces, demonstrates the hills and valleys of mountain making.

Why:

Great broken pieces of the Earth's crust, called plates, float on the layer beneath the surface called the mantle. This is similar to cracked sheets of ice floating on water.

When these plates meet, pass, or bump into each other (the science of plate tectonics), great and forceful pressures are created. These tremendously strong forces can fold, bend, uplift, and break the Earth surfaces to form whole mountain ranges.

When the clay rope was pushed and forced inward from the ends, it made hills and valleys similar to mountain building. The hill shape, higher in the middle, is called an anticline formation. If the clay makes a wavy S-like pattern that dips in the center, it is a syncline formation. The North American Appalachian mountain chain is an example of mountains formed by folding.

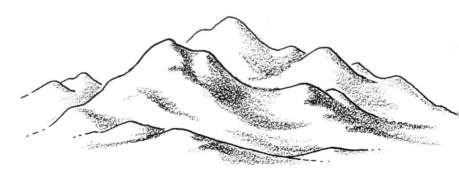

Tsunami: It'll Tide You Over

If you receive some money when you're very broke, we say it will "tide you over" or help you out until you get paid. But a *tsunami*, a Japanese term for great ocean waves, will really tide you over...tidal wave, that is!

In this activity, you can create conditions that will produce your own tsunami wave, then you'll understand much better how they are formed and the forces and dangers that these giant tidal waves produce. This is a great experiment for a hot summer day because it's likely you will get very wet! So either wear your old clothes, or be very careful.

You need:
a deep aluminum or other   water
   baking pan   2 blocks of wood

What to do:
Fill the pan with water and place the blocks of wood in the bottom of the pan so they are completely below the surface of the water. The object of this experiment is to rapidly compress, or squeeze, the water between the blocks. So, take hold of the blocks and quickly bring them together. Do it again, and again. Continue this squeezing action until the blocks can no longer compress the water.

What happens:
The movement of the two blocks coming together rapidly under the water forces swells of water to the surface, where they form waves that splash over the sides of the pan.

Why:

The action of the blocks and the water in this experiment is similar to the conditions in the ocean depths that produce tsunami tidal waves. Great earthquakes and volcanic forces on the ocean floor cause large amounts of ocean water to be compressed, or squeezed together, and pushed to the surface. There, great walls of water are formed and threaten nearby coastal cities. These great tidal waves sometimes reach heights of 50–100 feet (15–30 m). Because they form so suddenly and without warning, they are extremely dangerous and often kill many people.

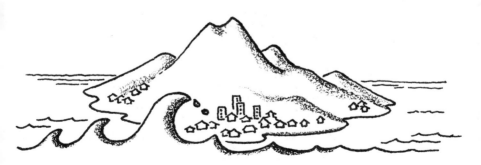

Photoplay: Say Cheese!

Light energy from the sun is so important that without it there would be *no life on Earth*! Still, we can put this great energy to use right now by making a simple but exciting pinhole camera. It uses light rays from the nearest star—our own sun, Sol—and is one of our brightest ideas. So get ready, set, and smile and say "Cheese!"

You need:

shoebox with lid
black tempera paint
   (available from craft,
   hobby, drug, or
   variety store)
brush

piece of wax or tracing
   paper cut into a
   3 x 5 inch (7 x 12 cm)
   rectangle
scissors
tape

What to do:

Prepare your camera by painting the insides and the lid of the shoebox with black paint.

Cut a 2 x 4 inch (5 x 10 cm) opening in the middle of one end panel of the shoebox and tape a larger piece of tracing or wax paper over it. You should now have a screen on one side of your pinhole camera.

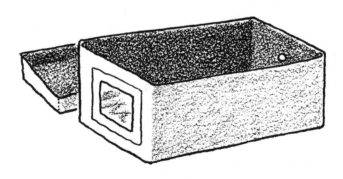

At the other end of the box, again in the middle of the panel, carefully punch a small ³/₈ inch (½ cm) hole in the side with the scissors.

Now you are ready for action. Take your camera outside, find a sunny location, and place something—a friend, toy, or object—in front of it. Point the pinhole side of the camera towards the object, and keep the screen in position in front of you for viewing.

See "Picture Perfect: Watch the Birdie!" for picture-perfect images.

What happens:
When you aim the small opening of the pinhole camera at something, a fuzzy but noticeable, upside-down image of that object appears on the screen.

Why:
The image or picture on the pinhole camera is reversed because light normally travels only in straight lines. Light rays from the top part of the image are reflected to the bottom part of the screen while rays from the bottom part of the image fall on the top.

PICTURE PERFECT: WATCH THE BIRDIE!

To view a perfect picture, or image, through your pinhole camera, place a covering over your head. Wrap it around your head and the screen so that it is completely dark; no light is able to get in. (This may remind you of photographers long ago, with their big cameras on tripods, who had to cover their heads with large dark cloths attached to the cameras in order to take pictures.)

Find something or someone (your subject) in the light while looking through the screen held before your face. A part of a house or a person at sunset is a perfect image.

Move the camera away from your face, up or down, closer or farther until the object is in view. Take your time—it may take several trials to adjust your eyes to the dark, get enough light into the box, and find the object, but you will eventually succeed.

I Steam Cone

This is one treat you can't eat, but you and your friends will love. In this great party-trick experiment, you'll build a different type of volcano based on earth science and chemistry—it's definitely something to get all steamed up about! It's simple, easy and you won't need a lot of materials. So, what are you waiting for? Get going and dig in! (Caution! Throw away all chemical solutions and thoroughly wash out all containers when finished.)

You need:
strip of lightweight cardboard, 3 x 8 inches (8 x 20 cm)
small container (spice jar or vitamin bottle)
flat tray or pan
½ tablespoon quick-rising yeast
½ cup hydrogen peroxide
scissors
paper clip or tape
spoon

What to do:
With the card-
board strip,
form a cone
shape that
will fit over
the mouth of

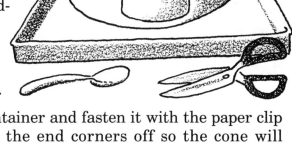

the small container and fasten it with the paper clip or tape. Cut the end corners off so the cone will stand upright in the tray or pan. Place the small bottle or jar in the tray and get ready for action.

The jar or bottle should be large enough to contain the hydrogen peroxide but fit under the cardboard cone or extend slightly above the cone's mouth. With the cone over the small container, pour in the hydrogen peroxide followed by the quick-rising yeast. Stir the mixture thoroughly. (If easier, you may place the cone over the bottle *after* stirring, but you must be quick!) Continue to stir the mixture, for best results, until the experiment is finished.

What happens:
The mixture of hydrogen peroxide and yeast causes foam, steam, and a hissing noise to come from the cardboard "volcano."

Why:
The ingredients placed in the container under the cone produced a chemical reaction, or change. It is called exothermic because, in addition to foaming, steaming, and hissing, heat is given off. If you touch the rim and sides of the container, or the stirring spoon especially if it is metal, you can feel this warmth.

In a real volcano, hot melted rock called magma, deep within the Earth, erupts or shoots through fissures or cracks. This moving rock, known as lava, sometimes flows from openings in the volcano's sides, or explosively shoots or blows out steam, smoke, ash, and rocks. Although your model volcano is small and simple, it does give you a good idea how a real volcano erupts.

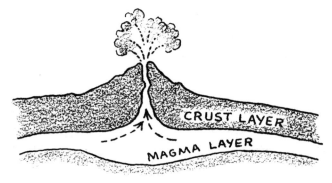

Hotwire Highjinks

Rocks inside the Earth can be changed due to the pressure and folding within the Earth, and such pressure creates heat. Do this experiment and see what we mean.

You need:

wire coat hanger   adult   candle

What to do:

Find an old coat hanger. Ask an adult to prepare for the experiment by unhooking the hanger or cutting it apart. Take the hanger and bend one section rapidly back and forth, in the same place, 30 to 50 times. Quickly, place the bent section against the candle. Don't touch the wire.

What happens:

The hanger wire has heated up. This warmth placed against the candle causes some melted grooves or ridges to appear on the candle.

Why:

Deep within the Earth certain rocks, called metamorphic, are caused by the constant folding of the Earth. This causes heat and changes the composition of the rocks, their make-up. Marble and quartz are examples of metamorphic rock.

In our experiment, the rapid and constant bending of the coat hanger caused heat that changed or partly melted the wax, the same way that pressure and heat within the Earth melts and changes rocks.

An Earth-Shattering Experience

Limestone caves are hollowed out by rainwater that is slightly acid and, over thousands of years, has gradually, eventually, and greatly dissolved away the soft rock.

You need:
a piece of chalk
½ cup vinegar
small jar

What to do:
Place the piece of chalk in the jar with the vinegar for five minutes.

What happens:
The chalk instantly dissolves in the vinegar solution, or acetic acid.

Why:
School chalk is a form of limestone, or calcium carbonate. It is made up of small bits of seashells and the mineral calcite and is similar to the soft rock caves of limestone. These caves have been formed by the rock that has been dissolved by the acids in rainwater, similar to the chalk that is dissolved by  the vinegar (acetic acid). England's famed White Cliffs of Dover are made up of great sheets of chalk, a form of calcium carbonate.

Shell Shock

Replace the chalk with a few seashells, another form of calcium carbonate and limestone, and see how fast or completely they dissolve.

You need:
some seashells
2 small jars
½ cup vinegar
½ cup water
newspaper
spoon

What to do:
Place some seashells in a jar with the vinegar, and a few in a jar with the water (as the control, for comparison). Let the shells sit in the solutions for three to four days.

Remove the shells from the jars, place them on newspaper on a counter or worktable, and carefully try to break them with the spoon.

What happens:
The shells from the jar of water remain the same, while the shells placed in the vinegar should break and crumble quite easily. They will also be covered with a white chalky substance (calcium carbonate).

Why:
The shells in the water are the experiment's control, to be compared against those in the other jar that were affected by the vinegar. Again, there is acid in rainwater as there is acid in vinegar. Acid will dissolve calcium carbonate whether it is in the form of cave rock, chalk, or shells. In some areas of the world, the rain is as acid as vinegar.

Glacier Melt

You can learn a lot about glaciers by making a model of one. It's best to do this outside. Adult help may be needed.

You need:

a small cup or yogurt container
sand
small rocks or pebbles
water
freezer

piece of board, to make incline or slant
hammer and nail
a thick rubber band
watch

What to do:

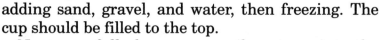

Place a one-inch (2.5 cm) layer of sand and gravel in the cup, followed by a few inches of water. Place it in the freezer. When frozen solid, repeat the process, adding sand, gravel, and water, then freezing. The cup should be filled to the top.

Next, carefully hammer a nail partway into the middle of one end of the board. Place that end against something immovable to form an incline or slant. Now you are ready.

Remove your model glacier from the freezer. Warm the sides of the container under warm tap water just enough to get your model glacier to slide out when tapped. With the rock/sand-side down, place the glacier at the top of the incline and fasten the rubber band around its middle and around the nail. How long will it take your glacier to melt, move, and leave rock and sand deposits? Time it.

What happens:

Depending on the weather, melting should begin immediately, even on cooler days. Rock and sand deposits will fall off in clumps, some will slide down the board, while other separate bits and pieces will form along the board surface in strange patterns, much like moraine or glacial matter.

Why:

Glaciers are large masses of ice that move down mountainsides and valleys cutting further gouges out of the rock and soil. Deposits from glacier movements can be found in such places as the Arctic, Antarctica, Finland, and Greenland.

These giant masses of ice would not move at all if it weren't for the great pressures they also exert. The force of these pressures causes periods of heating, and melting. The ice refreezes, but just enough thawing occurs to cause the slipping movement.

As glaciers move, they break off and pick up tons of rock and soil and deposit it someplace else. The unusual rock formations or deposits left behind are called moraine. Like the real thing, our miniature-glacier experiment shows us how and why those rock and sand deposits are so unusual and often unevenly placed.

SNOW...ER! ICEBALLS!

This simple activity can help you to understand how huge glaciers are formed.

Glaciers happen when snow becomes compacted, or packed tightly together. During the wintertime, when snow is on the ground, go outside and get some. At other times, ask someone to make you some shaved ice in a blender or food processor, so that you can learn about glaciers.

Compact or squeeze the snow or shaved ice into a tight ball (notice how solid it becomes). Let it melt a little and then put it into the freezer for about 30 to 60 minutes. The removed chunk of snow will be changed into a solid ball of ice.

Think of all the snow that falls in the mountains, day after day, compressing the snow underneath, and what happens to that snow, and you can imagine how the great glaciers are formed.

WORLD

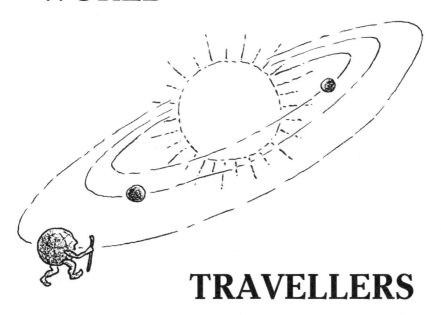

TRAVELLERS

While the Earth's surface, or crust, is always changing, and our planet continues to move through space, time and the seasons go through constant cycles—day after day, year after year.

While reading about the Earth and its place in the solar system of planets is good, doing simple experiments that help you to understand time and space and why things happen as they do is even better.

So, gather up your materials and get ready to do some simple, interesting, and timely experiments. Definitely, a fun time will be had by all!

Stick Around

Make a simple sundial or sun clock and watch the shadow from its stick or rod, called gnomon, move around on the ground to tell time. The angle of the shadow produced by the sun will change as the Earth rotates, or spins, and changes from day to night.

So stick around to watch the shadows, and the time, change—it's time well spent!

You need:
a stick
stones or other markers
a sunny day
pencil and paper

What to do:
Find a sunny location in your yard and push the stick into the ground. On the hour, mark the time of day on the piece of paper and place a stone or marker on the spot where the shadow strikes the ground. Again, one hour later, record the time and mark the shadow with a stone or marker. Continue these steps until you have a completed and marked (calibrated) a sun clock.

What happens:
The shadow cast by the sun on the gnomon, or stick, will change angle and length as the sun moves from east to west in the sky.

Why:
Although the sun appears to be moving from east to west, it is really the Earth that is moving or revolving around the sun. Besides orbiting, or circling, the sun, the Earth also spins on its axis, or turns like a

top. It is this spinning, or rotating, in relation to the sun that allows us to be able to record time, and night and day. Where the sun casts the gnomon's shadow at a certain time of the day, it casts the same shadow the next day, and the day after that, as long as the Earth turns.

In the morning, the shadow will be long and narrow and will point to the west. At noon, when the sun appears at its highest point, the shadow will be short and will point north (in the northern hemisphere, but south in the southern hemisphere). In the afternoon, the shadow will be directed toward the east.

Time on My Hands

It is important to check and double-check all experiments to make certain that the results come from your hypothesis, or scientific guess, as to what will happen. If you are not careful, other causes or variables could affect the results you get.

To make certain that shadows always perform the same, do the "Stick Around" experiment again, but this time use the shadow of a hand-held pencil gnomon to compare with your shadow stick. This is a "controlled" experiment, comparing the shadow of the short pencil rod to the longer stick rod.

You need:
2 pencils
a piece of clay
pencil and
    paper

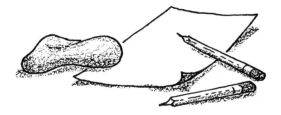

What to do:
Flatten the piece of clay into a disc about two inches (5 cm) in diameter for a base. Push the pointed end of the pencil into the clay. You now have a simple hand gnomon. Position this gnomon, aligning and adjusting it according to the shadow cast by the stick version. Write down the exact position in which you place the hand rod next to the shadow stick and keep it in that position every time you do a reading or test the experiment. (Example: Place the clay base of the hand rod next to the ground stick and align or adjust the shadows of the two gnomon rods so that they are parallel, or next to, one another.)

Is the shadow in alignment or in the same position as the other? Mark the base of your hand gnomon with the side of the other pencil. Put pressure on the clay, marking where the shadow strikes it. Continue to compare the shadow of the pencil gnomon with the rod in the ground. Mark each, with stones or markings, every hour. What are the similarities or differences? Do the shadows change position equally? Do the shadows of the rods grow longer or shorter? If so, when?

Spotlight Time Machine

The sun is a sort of time machine, as this simple experiment will show.

You need:
a cardboard square, about
   4 x 4 inches (10 x 10 cm)
a sunny window
watch or clock
scissors
tape
2 pieces of paper
pencil

What to do:
Cut a two centimetre hole in the middle of the cardboard square and tape it on a south-facing window. Tape the square in a position on the window so the spot of sunlight will shine on a clear area of floor. Place the paper where the spot of sunlight hits the floor so that it lands on the paper.

   Draw a circle around the spot of light and write the time next to it. Continue to watch the sunlight spots for thirty to forty minutes. Use more paper and record the times and movement as you watch.

What happens:
The spots of light move from left to right and change their positions as the time changes.

Why:
The Earth rotates, or turns, from west to east every 24 hours while it travels around the sun. This movement of the Earth causes the spot of sunlight to move across one paper and to the next as your part of the Earth moves from sunrise to sunset.

SEASON TICKETS

Would the spots of sun fall in the same location at 8:00 a.m. in the summer as they would at 8:00 a.m. in the winter?

If long-term experiments, those that take a while, don't bother you and you can find an unbusy room with a south-facing window, try it! An undisturbed bedroom with plenty of floor space would be an excellent place to do this experiment.

Repeat "Spotlight Time Machine," but this time replace the paper with cut-out paper tickets about 4 inches (10 cm) square.

Tape them to the floor, over the spots, at the same time at different times of the year. Example: 8:00 a.m., October 1, 1996, and 8:00 a.m., December 22, 1996.

Is there a difference in the positions of the spots from one season to the next? Record your results.

Skylight Direction-Finder

Early explorers and sailors used a simple direction-finder called an astrolabe to find their location on the open sea. You can make your own astrolabe and find your location on the Earth with a few easy-to-find and inexpensive materials.

You need:

protractor (from stationery store
   or with school supplies)
a weight (washer, screw,
   paper clip, etc.)
new pencil

string
scissors

What to do:
Tie a piece of string from the middle of the flat side of the protractor. The string should extend a little beyond it. Now tie the weight to the end of the string and place the flat side of the protractor next to the pencil and tie it to it with two additional short pieces of string.

You are now ready to use your astrolabe. On a clear, starry night, point your instrument and center it on the North Star, sometimes called the Pole Star. The weighted string will drop next to the side of the protractor and will show the degree or the number of your latitude. Be patient and do this several times to get an accurate reading.

What happens:
Your instrument, when pointed to the North Star, will help you find your latitude on earth. Latitude is

36

a series of imaginary side-to-side Earth lines that tell you in numbers where a certain place on Earth is located.

Why:
The protractor in your astrolabe is a half-circled instrument used to measure angles. It is marked in units of ten called degrees. When you pointed your instrument to the North Star, the weighted string aligned itself to the unit angle on the protractor. This, in turn, gave you your exact location on Earth or your latitude.

See "How to Find the North Star."

HOW TO FIND THE NORTH STAR

The North Star, also called the Pole Star, is seen only in the northern part of the sky. This star seems almost fixed in place because of its position above the North Pole. Like a clock, it also appears to change position from hour to hour and season to season. It appears very faint because it is over 400 light-years away.

The North Star can be found opposite the constellation Ursa Major, commonly known as the Big Dipper. It is seen as a cup with a long handle on it, if you were to draw imaginary lines from each of the seven stars that make up the pattern (as in dot-to-dot). The stars furthest from the handle, which make up the cup, point directly to the North Star.

THE SOUTHERN CROSS

Do people in the southern hemisphere, or the southern half of the world, also see the stars visible from the northern half of the Earth?

If you live in the northern hemisphere, you cannot see all the stars in the sky over the southern half of the world. This is true the opposite way in the southern half.

If, however, you live on or near the imaginary line that circles the middle of the Earth, known as the equator, you can see all the stars of both halves of the Earth.

While people in the northern hemisphere can find Polaris, the Pole or North Star, which appears to be fixed directly above the North Pole, star gazers in the southern hemisphere have no such marker.

The Southern Cross, in the southern hemisphere, is a constellation or group of stars made up of four crossed stars and other stars, two of which, like the north's Big Dipper, point to the South Pole. The Southern Cross, however, is not a good locator of its pole. It does not always appear as a cross and often it is hard to find and see. Too, the southern hemisphere has no visible "pole" star to mark it in the sky.

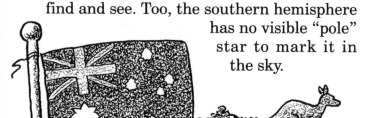

Meteor Burnout: Truth or Friction?

Meteors are small rocklike chunks, most probably broken fragments of comets or asteroids, tumbling rapidly around the sun in outer space. As they enter and pass through the Earth's atmosphere, they burn up. Now you can try this simple experiment and discover how this happens.

You need:
2-litre soda bottle filled with water
½ of a seltzer tablet

What to do:
Drop the tablet half into the bottle of water and watch what happens as it falls or floats down to the bottom.

What happens:
The tablet dissolves or breaks up into many small pieces or fragments that disappear as it journeys to the bottom of the bottle.

Why:
The water represents the Earth's atmosphere and the seltzer tablet, the meteor.

Like a meteor, the tablet breaks up into many small fragments as it drops to the bottom of the bottle (Earth's surface). Unlike the tablet, the meteor rushes through outer space at such great speeds that friction, or the rubbing force of its surface against the Earth's atmosphere, causes the space rock to heat up and the white-hot fragment to break up and explode into cosmic space dust.

Most meteors are no bigger than small stones, but every so often a few larger chunks make their way to the surface of the Earth as meteorites.

Starry-Eyed

Make a star-lit light box and learn about the constellations, groups of fixed clusters of stars. It's fun and it's easy, and you and your friends will be headed for stardom.

You need:

oatmeal box with lid	pencil
(several lids are best)	dark room
flashlight	
nail	

What to do:

Punch "star" holes with the nail in the box lid. Follow your favorite star-chart patterns from astronomy books, or from looking overhead at the night sky where you live.

Depending on the size of the box and lid, you should be able to punch in a large pattern or two or more smaller constellations on a lid. Collect and use other lids, or make your own, for new constellations to enjoy. By rotating a lid or turning the box, you can even make the constellations move.

Now, press the narrow end or handle of the flashlight against the center of the other end of the box and draw a circle around it. When finished, cut a hole in it and fit the flashlight into the box. (This can be done by removing the lid and pushing the handle through the hole.)

You are now ready to dazzle your friends with your new star-lit light box . Take it into a dark room and point it, with the light on, at the ceiling or wall and enjoy the starry-eyed show.

What happens:
Your light box projects groups of small, star-like, light spots on the ceiling or wall.

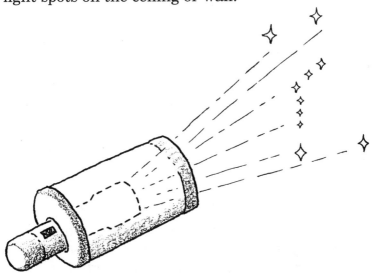

Why:
Different constellations seem to sweep across the sky, and are seen only at certain times of the year. From week to week, and at the same time every night, their positions change, as they move a little farther to the west.

The orbit or path the Earth takes around the sun, and its position at certain times of the year, determines whether you can see certain constellations or not. In winter, the summer constellations are blocked out by the sun's light while the winter constellations are blocked out in the summer.

Parallax Puzzle

Scientists can calculate, or figure out mathematically, the distances of different stars from the Earth.

As we look at the stars, we understand from what we have learned that the stars stay in an exact position and are very far away. But what our eyes and our brains tell us may not be totally correct.

If we walk past a house, the house does not move but the position of it does because the angle we view it from after passing it is different. This is parallax, and this simple experiment will show us how parallax works.

You need:
a pencil   your two eyes

What to do:
Hold the pencil out vertically, or straight up and down, right before your eyes. Now, close your left eye, then quickly open it and close your right eye. Do it again. Continue to rapidly close first one eye and then the other and observe what happens to the pencil before you.

What happens:
The pencil jumps, moving from side to side! Where is it *really*? How can you know?
Going on...

Moving Picture

Do the parallax experiment again the same way, but this time, as you view the close-up pencil, look also at a distant object, such as a lamp or a table, in the background. What do you see? Does the position of one object change more than the other?

What happens:
Although the pencil appears to shift position, or move, from one side to the other as it did before, the distant object did not.

Why:
The pencil did not really move but the angle of viewing did. The angle between the pencil and your eye changed and so the pencil's position did, but only according to your eyes and your brain. This shift difference is called parallax, and it is the key to finding distance. The closer an object is, the more it will seem to shift, while objects that are farther away do not. In the same way, parallax makes closer stars seem to move; and those very far away seem fixed.

To see the shifting of stars, astronomers (scientists who study and observe the universe from our solar system to the farthest galaxies) measure the different positions of stars at two different times of the year as the Earth moves in orbit around the sun. In that way they can calculate the stars' exact distance from the Earth.

Track Star

Our sun, a star, is a giant ball of hydrogen gas many million of miles away. Yet it is possible to learn something about the sun from tracking, or following, the wavelengths of light coming from it. This can be done in a simple way.

You need:

glass tumbler half-
   filled with water

sheet of paper
sunny outdoor location

What to do:

Find a place outside in full sunlight for your experiment. Place the sheet of paper on a table or lay it on the ground where the experiment is to be done.

Now, hold the glass with the water firmly between the thumb and a finger over the sheet of paper. The glass should be held about three to four inches (7–10 cm) above the paper. Do not hold the glass in the usual way, around the glass. It's important that you hold the glass so that your hand does not block the sides of the glass.

Be careful not to drop or break the glass or you can cut yourself.

Move the glass up and down and slant it slightly, focusing the light on the paper until a clear colorful pattern is noticed.

What happens:
The glass of water acts as a prism and casts a rainbow on the paper.

Why:
A glass of water is able to act as a prism, or something that can change the direction of light so the bands of color in it can be seen and studied. White light is really a combination of many colors.

When a wavelength of light is split and changed by the glass of water, color occurs. Light from the sun shows many colors. Astronomers can tell what elements or gases make up a star by studying the bands or spectrums of the light it gives off.

GLASSIFY

Do different types of glass make better prisms for casting rainbow patterns on paper?

Do the same experiment as above, but instead use a different-size glass, then one with a different shape. What about colored glass? Will glasses made of colored glass refract, or split, light into colors, too? Will a full glass of water work better than half. a glass?

Do a variety of experiments and write down your observations and results. When you're finished, you'll know just what glass works best.

Highly Focused

The seasons of the year depend on the tilt of the Earth and the concentration of sunlight at different times of the year in the northern and southern hemispheres. This simple experiment explains it all.

You need:

watch or clock
flashlight or electric
    lantern
thermometer

can or other support
    for thermometer
paper and pencil

What to do:

Record the temperature on the thermometer. Run the thermometer under warm or cool water to get the temperature where you want it, so that it is easy for you to record and calculate.

Prop the thermometer, glass side outward, against the lighted end of the flashlight. Leave it in that position, timing it with a clock or watch, for three minutes. Record the final temperature.

After the first reading against the flashlight, hold the thermometer under cool water until the temperature returns to what it was at the beginning of the first trial.

Now, lean the thermometer against a support to hold it upright and shine the flashlight on it from a set distance away, say one foot or 30 centimetres. Again, record the reading after three minutes.

What happens:

The thermometer leaning against the flashlight and in direct contact with the light, so it was more concentrated or had greater strength, registered a few degrees warmer. No noticeable change was seen when the light was shined on the thermometer from a short distance.

Why:

The concentration of light on various parts of the Earth, at any one time of year, is similar to the concentration of light in our experiment.

The greater an area covered by light, the less the temperature. In our experiment, the thermometer that was farther away from the light was not affected by it as much, if at all.

The northern hemisphere, or upper half of the Earth, which is tilted away from the sun in December receives a greater spread of light, while the southern hemisphere, or lower half of the Earth, is tilted towards the sun at that time and receives stronger, more concentrated light. This explains why, in December, it is winter in New York City and summer in Sydney, Australia.

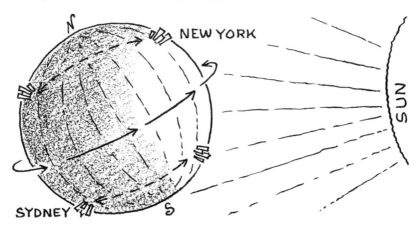

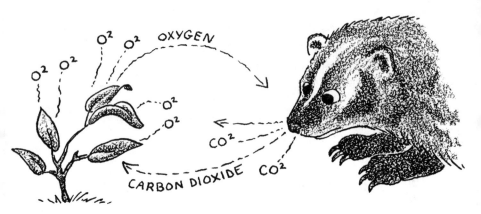

LEAFY LESSONS

We could not live without plants. Consider these facts:
• plants, humans, and animals keep the Earth's atmosphere balanced
• plants, through photosynthesis, make their own food, while taking in carbon dioxide and giving off oxygen
• animals and humans need oxygen and breathe out carbon dioxide
• animals and humans get much-needed sugars and starches from eating plants
• millions of tons of water are released into the air every day by plants (transpiration)
• some scientists believe that the loss of trees and an increase in human and animal breathing, or respiration, can raise the amount of carbon dioxide in the atmosphere and cause the Earth to warm up.

The experiments in this chapter will answer your questions about plant growth, but more important, they'll show you why these living things are so very important to our lives.

Oxygen Leaves

Oxygen leaves? That's right, oxygen leaves! Leaves what? Confused? Try this experiment and you won't be. You'll also learn about two important plant words—stomata and photosynthesis.

You need:
small, clear, wide-mouth jar
leaf
magnifying hand lens

What to do:
Fill the container with water and drop the leaf into it. Place it in a sunny location outdoors or on a windowsill. Leave it there in the sun for at least one hour, or until the container feels warm. With the hand lens, look at what happens inside the container.

What happens:
Thousands of tiny bubbles appear on the surface of the leaf and the inside of the container.

Why:
The bubbles are formed by the oxygen gas given off by the leaf.

A plant needs certain elements and sunlight to make its own food. This process is called photosynthesis. "Photo" and "synthesis" mean "light" and "putting together." When water, air, chlorophyll (which causes the green coloring), and sunlight are put together in a certain way by the plant, it makes its own food. If any one of these elements is missing, a plant cannot live.

Carbon dioxide, a gas, enters the tiny, pin-like holes in the underside of a leaf, called stomata. The plant uses sunlight and chlorophyll and, combined with water and carbon dioxide, turns these elements into the food it needs. The food is actually a form of sugar that is eventually turned into starch. Oxygen is given off as a waste product. Now, you know why you saw the

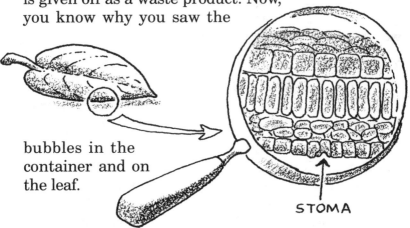

bubbles in the container and on the leaf.

STOMA

DON'T LEAF ME ALONE!

Now see what happens if you do the experiment but place the container in the shade.

Does it matter if a leaf is in the sunlight *before* you test it? Try placing a leaf from outdoors in the sunlight into one container of water and a second leaf not from the sunny outdoors, e.g. an indoor plant, into another container. Put both containers outdoors in the sunlight. Is there any difference?

Try this experiment indoors now. Do bubbles appear on the leaf or container?

Remember always to keep good notes and records and to log, or write down, all your observations and results.

Phototropism: Waiting
for the Weedends

Plants will always grow towards the sun. They will turn upwards, even if they are turned on their sides.

You need:

2 small pieces of plastic, about 3 inches (8 cm) square
2 rubber bands
a small shallow container (from frozen food package or meat tray)
pencil and paper

1 paper napkin
scissors
1 medium-size weed with well-developed leaves and stems and a good root system

What to do:

Take two flat squares of plastic. You might be able to cut them from a plastic container. Fold the napkin to fit onto one of the plastic squares and arrange the roots of the plant on top of it.

Place the second square of plastic on top, like a sandwich, making sure that the stems and leaves of the plant are outside of the plastic sandwich. Finally, wrap the rubber bands tightly around the squares to keep everything in place.

Place the "weed sandwich," propped on its side in a shallow, water-filled container, near a sunny south-facing window. Make certain that there is always a half to an inch (1–2 cm) of water in the bottom of the container at all times. Make drawings showing the position of your plant each day. Be patient. Wait at least four to five days for results.

What happens:
The leaves and stems of the weed grow upwards, towards the sun, even though the weed was placed on its side.

Why:
A plant's leaves and stems will always grow towards the sun, no matter if they are placed sideways, or even upside down. This could involve moving, bending, and turning, so as to go towards the light. This process is known as phototropism.

DRYING THE INSIDES OF BOTTLES

Some experiments call for completely dry soda bottles. But how do you get the inside of a wet soda bottle dry? Just stuff a sheet or two of paper towelling inside the bottle and use a long-handled screwdriver, stick, or something else long and thin to press and stir the absorbent paper against the sides and bottom—and slide it up and out of the bottle when finished.

Perspiration or Transpiration: Don't Sweat It!

People sweat, or perspire, while plants transpire. A plant gives off water through its stomata, tiny holes located under its leafy surface. Now see what it's all about.

You need:

2 short clear soda or water bottles
a lump of clay (golf ball size)
magnifying hand lens

a broad leaf, or leaflet with stem, which has been exposed to the sun
pencil or nail

What to do:

Roll the clay between your hands to form a two-inch (4 cm) plug. The plug will need to reach about one inch (2 cm) into each bottle neck to hold one bottle vertically, upside down, above the other.

Using a nail or pencil, poke a hole in the plug and insert the stem of the leaf through it, being careful not to break the stem or crush the leaf. Now, gently, press the clay plug inward around the stem to seal it in. Fill a bottle with water and push the plug with the leaf in it into the top of the bottle. (The plug should rise above the bottle neck and the stem of the leaf must touch the water.)

Wipe any moisture from the plug or leaf and make certain that the plug itself is not touching the water—this could cause moisture to get into the bottle above and negate the experiment.

Carefully, turn the other bottle upside down on top, working the leaf into it, and the plug into place. Press the clay gently around any opening to seal it. After one hour, take the magnifying glass and observe your experiment closely.

What happens:
A small, but noticeable amount of moisture (areas with small water droplets and steamy haze) appears on the glass inside the "dry" upside-down bottle.

Why:
Transpiration in plants is much like a person sweating. A plant loses water vapor through holes or pores called stomata. Plants often obtain too much ground water through their roots and get rid of what is not needed through these holes.

Did you know all the world's water is always the same—none is ever lost? In other words, the Earth's waters are naturally recycled through rain, clouds, lakes, rivers, oceans, and especially by plant transpiration.

Although you may not see it, plants give off gallons of water each day. It's true! Leaves really do sweat, or transpire, so much that they release millions of tons of water vapor into the air every day. This is an Earth process we never consider, but without it, we could never live on this planet. Don't sweat it, but do be-*leaf*-it!

THREE LEAVES AND A MAYBE

In the last experiment, you used a broad leaf in a bottle and saw water droplets and steam form. But would the rate of transpiration be the same if different types of leaves were used?

Find more bottles and try different leaves—broad, narrow, leaflets, fern-like, etc. Place one set in the sun, another in the house. Is the amount of water seen in the "dry" bottle more, less, the same? What's your hypothesis?

Doing the same experiment but without the leaf can serve as a control, to show that other things are not producing the water droplets. To be a good earth scientist, keep good notes.

FEARS FOR THE RAIN FORESTS

Rain forests, where many tropical plants, trees, and animals live, are located in the warmer parts of our world. New products, foods, and medicines are being discovered every day in these rain forests. Trees and plants in the rain forests also take in great amounts of carbon dioxide gas, supplying oxygen to our Earth and keeping our air clean.

Now that you've seen what important work even small plants do, can you understand how destroying the rain forests, chopping down trees, can affect us?

For the Birds

Birdseed, radish seeds, onion seeds, or any other kind of seed will grow if placed on a water-soaked sponge.

You need:

sponge
shallow container
  (small frozen food tin)
seeds
water
hand lens

What to do:

Place the sponge in the dish or container with enough water in it to soak it. The sponge should rise above the water level and water should be added to the container from time to time as the water evaporates, to keep the sponge moist.

Sprinkle a small amount of seeds on the surface of the sponge and lightly pat them into it. Place the container of sponge-soaked seeds in a sunny location such as a windowsill. Check them in two to three days for some cracking and sprouting. You'll need your hand lens to see them. The seedlings

should be fully developed in five to seven days.

Why:

The dry seeds swell until they break open, when placed on the water-soaked sponge. The seeds germinate, or start to develop, on the sponge. The water softens the outer part of the seed called the

seed coat. At this stage, seeds need only food, water, and air to grow. The new plant uses the seed for food but eventually needs soil and sun to make its own food.

What now:

After the seeds have sprouted, gently scrape them off the sponge and let them fall into a container of potting soil or another gardening material such as vermiculite. Now, you have a quick and easy and fun way, because you can *see* them, to sprout seeds. (See "Planting Seeds and Seedlings" and "Waterbed" for instructions.)

Waterbed

Can plants be grown if you don't have any soil? Hydroponics is the science of growing plants without this needed element! How is it done? Does it really work? You'll find out, dirt free, in this important scientific investigation.

You need:

flower pots (with
   holes in bottom)
tray or shallow dish(es)
flower or vegetable seeds
water-absorbent plant
   material such as
   vermiculite, perlite, or
   peat moss

liquid or granular
   plant food
stones or broken
   pottery, for pot
   drainage
spray bottle

What to do:

Place the stones or broken pottery in the bottom of the flower pot, to cover the hole and provide drainage. Fill the remaining area with the planting material.

With the spray bottle, water the material well—it should be moist but not soaking wet. Now, lightly and evenly scatter the seed over the planting mater-

ial and press it down. If you have a lot of several kinds of seeds, it is best to use several pots for good spacing and better growth. Continue to keep the planting materials moist and place the pots in sunny, south-facing windows.

After the plants germinate, or sprout, water them with a combination of water and plant food. (See package directions on how to dilute the food with water.) Continue to water the plants whenever more moisture is needed, but do not over-water.

What happens:
The seeds grow into healthy seedlings or young plants without using any type of soil.

Why:
Plants need air, water, and light to grow but they don't necessarily need soil. Plants can be grown without soil by replacing the minerals they would normally get from the soil with liquid or dry plant food. Hydroponics, or growing plants without soil, may be the new way of growing plants for the future.

Phil O. Dandrun's Root Bare

A new brand of soda? No. Phil O. Dandrun is our nickname (an alias) for the philodendron. If you have this common house plant at home, try this experiment—if you don't, use a different plant. One way or another, leave it to us and you'll be raising a plant from a stem in no time.

You need:
stem section, about 4 inches (8 cm) long,
    of philodendron or
    other plant with
    side leaves
    removed
    (cut stem
    below
    leaf scar
    or bulge
    called the node)
jar or jars of water

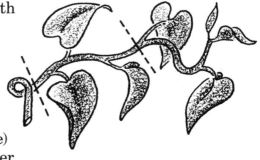

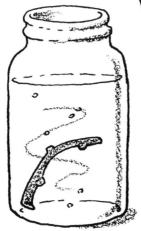

What to do:
Place the piece of philodendron in a jar of water. Make sure the water covers the spots where the leaves were pulled off, and patiently wait for roots to develop. It may take several weeks for good root growth.

What happens:
Long brown, thread-like roots form from the scarred areas where the leaves were removed.

Why:
Some plants can grow from roots, leaves, and stems. Philodendron sections or stems, called cuttings, that are placed in water will grow roots from the leaf areas around the bulge or node. Geraniums and other plants also do this, so if yours didn't work, don't give up. Try again.

Little Sprouts or Hothouse Tomatoes

We've learned that a greenhouse is a hot, closed space for growing plants. The Earth can have a sort of greenhouse effect, too, when gases from burning fossil fuels act as a lid and prevent heat from escaping into space.

But now, let's put this greenhouse idea to work for the fun of raising some sweet little sprouts that may even turn into hotshot tomatoes. There's nothing like tasting the juicy, sweet, delicious results of this successful experiment.

You need:

fresh tomatoes or pack-
  aged tomato seeds
pots or other containers
potting soil

spoon
plastic wrap
sunny windowsill
rubber bands

Note: Amount of materials needed depend on how many containers of seedlings you wish to grow.

What to do:

Prepare pots or containers with packed-down potting soil. Scoop seeds from fresh tomatoes or packaged seed and scatter them evenly over the soil, avoiding clumps of seeds; then cover them with a light and thin layer of soil.

Water well and cover each container with a piece of clear plastic wrap fastened with a rubber band. Place in a sunny windowsill and watch for little sprouts.

PLANTING SEEDS AND SEEDLINGS

Plant cuttings that have rooted, and seeds and seedlings, can be planted or moved into soil. Easy enough, but some rules must be followed.

For seeds to start growing and plants to grow well they need water—but not too much (they'll "drown" from lack of air) or too little (they'll dry out and die). Good light and soil with the proper minerals are also needed to boost growth.

First, plant seeds, seedlings, and new plants in clay or plastic pots with drainage holes in the bottom. (Egg and milk cartons with holes punched in them also make great planters.) Adding bits of broken clay pot, gravel, or stones also helps water drain. Use a good potting soil with equal parts of peat moss, sand, bark, wood, and a good nitrogen and iron content.

Containers should be nearly filled and plantings should be given "elbow room" away from other seeds, seedlings, or plants. In other words, don't crowd! A light covering of soil, about ¼ inch (1 cm), should cover seeds or seedling roots and be lightly compressed or pressed down. New plantings should be placed in a sunny, south-facing window, and soil should be kept moist. A small sprayer or medicine dropper might deliver just the right amount of water and help prevent over-watering.

If necessary, get an adult to help with your new plantings until you know what to do and how to do it to get the results you expect. Good luck and happy planting!

DIRTY WORDS: SOIL, SAND, HUMUS, AND MUD

The next time you fill a pot or planter with soil, think about its properties—how it looks, feels, and smells, what it's made of, and what lives in it. As you investigate, you'll discover bugs, leaves, pebbles, and small rocks, but that is just scratching the surface. You won't see the billions of microscopic plants and tiny animals that live there in the soil, too, but they are responsible for making up good, rich soils.

Bacteria, fungi, earthworms, gophers, moles, snakes, insects, and many other living things do the important work of decomposing or breaking down soils that help us to grow better and more nutritious crops.

In this chapter you'll learn about soil, sand, humus, and mud, and how dirt acts under certain conditions. Too, asking questions about *these* dirty words won't get you grounded!

DIRTY QUESTIONS CALL FOR DIRTY ANSWERS

What is soil made of? Is all soil the same? What's the difference? What kind of soil is best for most plants? How can I find out about it?

Read below, go on to the "Soil" box, and then do the "Earthshaking" experiment and you'll definitely dig up more dirt on the subject than you ever imagined was there.

What is known as soil or dirt is made up of broken rocks, minerals, the remains of dead plant and animal life, and germ-like plants called bacteria. These tiny, one-celled plants are too small to be seen, unless you look through a good microscope, but they are everywhere in the soil and do a big job.

Small animals, worms, oxygen, and water are also needed in the soil. With the help of the larger animal life, the bacteria use the air and water to break down and chemically change the soil. When their job is done, poor soil that was not good enough to grow anything becomes a nitrogen-rich fertile soil that can grow just about anything. (See also "Earthworms— Starring in 'A Real Dirty Deal!'")

SOIL, BY ANY OTHER NAME, IS STILL DIRT

Scientists have identified four different kinds of soil according to how it feels (texture) and what it contains. The types of soil are sand, silt, loam, and clay.

Sand is made up of broken shells and worn-down bits of rock and minerals such as quartz and basalt, a volcanic-like rock. Although all good soils need sand, too much of it can cause too much water to drain away from plant roots, leaving them to dry and shrivel up. Sand is found in deserts, on beaches, and along river bottoms. Larger grains are called gravel.

Silt, on the other hand, is a very fine-grained, sandy soil. Its parts are smaller than sand but larger than particles of clay.

Clay is a fine soil and it is much needed in all soils. Without it, soils fall apart and fertilizers are washed away. Too much clay in any soil, however, will cause problems with water drainage and eventually produce rotted roots in plants.

The best type of soil for most plants is loam. Loam is a mixture of clay, sand, and silt with enough humus (broken-down plant and animal matter) to make it rich and fertile.

SAND
SILT
CLAY

LOAM

Earthshaking Discovery: It's Sedimentary!

The sediment, or different types of soil particles, and how they float and settle, are bound to be unusual and interesting. Just shake up these soil-shakers and watch out.

You need:

jar(s) with lids (depending on how many soil samples you wish to test)
½ cup each soil sample from different locations and depths (topsoil or upper soil vs. subsoil or deeper soil)

water
magnifying hand lens
paper and pencil

What to do:

Fill the jar with the dirt and add water. The jar should be about three-fourths filled. Screw the lid on the jar tightly and shake well. Repeat the procedure with any other soil samples to be tested.

Be patient and wait about two hours for the soil to settle. (You could sit and watch, but you don't have to.) Then, with the hand lens, observe what happened to the soil samples. Draw a picture of the settled sediment in each jar.

What happens:
The soil in the jar(s) settles into bands or layers depending on the content of the soil.

Why:
In sandy mixtures, heavier, rock-like particles settle first, followed by light-colored silty, sand-like grains.

In most loamy, gardening topsoils, the heavier, gravel mix settles to the bottom, while the dark-colored, lighter-in-weight humus floats to the top of the jar. As you can see, this is a good test for determining good, loamy, rich soils.

What now:
Collect soil samples while out of state during long car trips and vacations and discover how much humus and kinds of soil are in each sample. If you're careful, you can keep your dirty secret and still find out what kind of dirty state you're in.

Air Condition

What if you can't *see* what condition the soil is in?

You need:

a small jar   a cup of boiled and cooled water
½ cup dirt   magnifying hand lens

What to do:

Place the soil sample in the jar. Pour the cooled, boiled water slowly onto the soil and watch closely.

What happens:

Air bubbles appear and circle the top surface of the soil.

Why:

All dry soil contains air trapped in and around the particles. The bubbles that rise from the soil's surface are formed by air forced from the soil by the water.

Water also normally contains its own air, which is why, for this experiment, it is necessary to use boiled and cooled water. During the boiling, the heated air in the water is boiled away. This experiment, then, reveals that it is air from the soil that causes the bubbles, and not air in the water.

Sand Trap

Quicksand is a thick body of sand grains mixed with water that appears as a dry hard surface. It may look solid, as if it can be walked on, so it can be unexpectedly dangerous because it really cannot support much weight. People have been known to be swallowed up in quicksand.

In this experiment, you'll make a type of quicksand goop that will magically and surprisingly support your hand one minute but not the next.

You need:
large bowl
sheet of newspaper
1¼ cup cornstarch
1 cup water
2 tablespoons ground
  coffee
spoon

What to do:
On newspaper, because this can be messy, place the cornstarch and water in the bowl and mix with the spoon until the ingredients look like paste. The cornstarch mixture naturally will be hard to stir and will stick to the bottom of the bowl. This is to be expected. Next, lightly and evenly sprinkle the ground coffee on the top of the mixture to give it a dry and even look.

Now the fun begins. Make a fist and lightly pound on the surface. Notice what happens and how it feels. Next, lightly push your fingers downward into the mixture.

What happens:
When you used your fist to hit the surface of the mixture, it appeared to hit the surface only and seemed to be mysteriously and magically stopped from going any further. But when you placed your

fingers or hand in the mixture, they easily and readily slid into the bottom of the bowl.

Why:
The molecules of quicksand goop behave much like the real thing. Unlike water molecules, the goop's molecules are larger, swell and hook together, and seem to act more like a solid than a liquid. In addition, the coffee grains give the mixture a deceptively smooth and dry look, much like the real sand.

A Down-to-Earth Water Filter

Have you ever wondered how water is cleaned before it reaches your home? How about making a simple water filtration system that will answer many questions. You can get lots of down-to-earth information as you test it.

Remember, though, that however good a job you think you have done, the water from this experiment should *never* be drunk. The experiment will give you a good idea how water filters work, but it is still not a real water treatment plant, and just a few drops of "bad" water can make you very sick.

It will be best to perform this activity outside since it can be messy. Too, the dirt you need to use should be easy to find nearby.

You need:

a medium-size flowerpot
  (or waxed carton with
  holes punched in the
  bottom)
coffee filter, thin cloth,
  or paper towel
2-litre soda bottle with cap
2 shallow trays or containers

gravel/small stones
sand
funnel
dirt
water

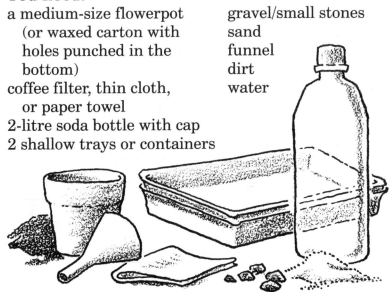

Note: Clean sand and gravel are available in small bags in garden or variety stores.

What to do:
Place the piece of filter or towelling in the bottom of the pot or carton. Fill the bottom of it with gravel or small stones, to a depth of about 2 inches (4 cm). Pour sand into the container until it is about three-quarters full.

Using a funnel, place about one cup of dirt into the 2-litre soda bottle and fill it with water. Screw the cap on and shake the bottle thoroughly.

Pour some of the muddy water into one of the shallow containers. This will be the control or test container, to compare the filtered water against the original sample. Place your filter system in the other container and pour some muddy water into the top of it. Watch the water as it filters through

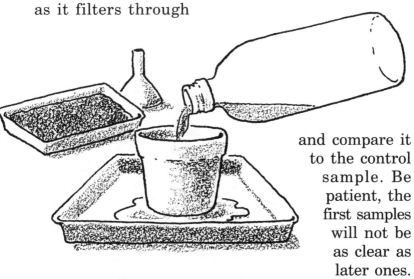

and compare it to the control sample. Be patient, the first samples will not be as clear as later ones.

Repeat this procedure several times until the water comes through fairly clear. Continue to compare these samples with the water in the control pan.

What happens:

The first water that trickles out of your filter system will still be fairly dirty. However, as you continue to pour the water back into and through your system, it begins to get clearer. Although the water gets cleaner, there will likely still be a certain amount of sediment that remains.

Why:

Although there are similarities, your simple water filter system is not like a city's large water treatment plant. In the city system, water is sprayed into the air to release unwanted gases, and substances are added to clump together dirt particles suspended in the water so that they can be filtered out.

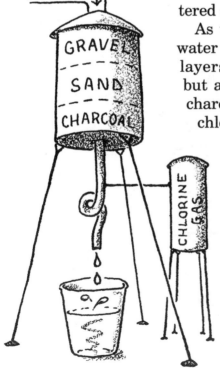

As with your system, the water is also passed through layers of sand and gravel, but also through a layer of charcoal. The water is then chlorinated. The chlorine gas kills bacteria that may be present in the water. Cleaning water is a big job.

Now you know something of how water filter systems work, and why you wouldn't want to drink any water you "clean" yourself.

Bubble Blowers

Find some porous rock (rocks that are lightweight, with holes or spaces in them) and place them in a pan filled with water for a rocky bubble-blower show.

You need:

porous rock or broken pieces of brick or pottery

shallow tray or tin
magnifying hand lens

What to do:

Place the rocks in the tin and cover them with water. Observe what happens using the hand lens.

What happens:

Streams of bubbles flow from the rocks. The more porous the rock, the more bubbles you will see. Depending on the weight of the rocks and the force of the air escaping from them, the rocks might move slightly, rock back and forth, or bounce and rattle against the pan.

Why:

Oxygen is present, even in the rocks. Air bubbles flow from the spaces in the minerals making up the rocks and rise to the water's surface.

Holding Pattern

Which soils hold the most water, the least, or just the right amount?

A soil is called "permeable" when it allows water to pass through it. Which soils are the most permeable? For this dirty job experiment, it will be best to set up work space outside where you can just let yourself dig in!

You need:

equal amounts of several
    different soil samples
    such as clay, sand,
    potting or rich loam
    gardening soil
paper cups, small pots, or
    waxed carton
    bottoms

small containers
water
paper and pencil
sharp pencil or nail

What to do:

With a nail or pencil point, punch about six holes in the bottom of cartons or paper cups. If pots are used, make certain they have holes for drainage. Fill each container about half-full with each soil sample to be tested. Pour a half-cup of water into each cup or pot of soil to be tested and place the container under each to catch the water. Pour the water that drains into each container into a measuring cup. Record the type of soil tested and how much water the soil held. Repeat this step with other soils and again measure the amount of water.

What happens:
There will be noticeably more soil and water in the bottom of some cups than in other. In some samples, water will drain faster while in others it will be much slower.

Why:
Clay soils retain, or keep, too much water, while sandy soils drain too quickly. Too much water around tender roots can cause rotting, while too little water can allow the roots to dry out and shrivel up. Soils with much humus, or decomposed, broken-down plant and animal matter, are best for most plants. It retains just enough water for healthy plant growth while stimulating the roots. Some plants, however, still do well or better in other kinds of soil.

Sand-Casting

The sea wears away coastal shorelines and rebuilds new sand formations. In this simple experiment, we'll see how the Earth is constantly being worn away, eroded, and how the process of erosion steadily changes the different shapes and formations on the Earth's surface.

You need:
aluminum baking pan or flat container   water
sand or very fine soil

What to do:
Pile the sand at one end of the tin and firmly pat it down. For the purposes of the experiment, this will represent the sandy beach or shore. Pour some water into the middle of the pan until part of the shore is slightly covered. At first gently, then increasingly faster, slide the pan back and forth until small waves are formed that roll up and unto the shore so that the sand shifts or moves.

What happens:
The action of the waves in the container gradually changes the shape of the shore, moving the sand down the beach and into the water.

Why:
All the seas of the Earth are always changing the land they meet. Some wear away or carve out great rocky areas of land while others take away great sections of sand, depositing it elsewhere. This gradual but persistent action of water against land is called erosion.

Playing Dirty, or Groovy Soil Boxes

If you liked playing in mud when you were small and even now like making clay pots in ceramics, you're going to love this experiment. Do it outside wearing old clothes because you can get pretty dirty if you're not careful.

You need:

three 1-quart (946 mL)
   milk cartons
garden soil
a measuring cup
flat pan or container

scissors
paper, pencil
short stick
water

What to do:

Cut away one side of each quart carton, the one away from the opening or spout. (The spout should rest on the ground.) Pack each opened milk carton with the same amount of soil.

Wet the soil in each box thoroughly (use a garden hose if you wish) and mix it up. If the soil is too wet, put in some dry garden soil and mix it thoroughly by hand. (I told you you'd like this!)

Next, pack down the muddy soil in each container to form a hill or slant with the high side towards the unopened end of the container. With your hands or a stick, form sideways or horizontal gullies or ridges in one carton, steps in a second carton, and leave the third "hill" alone.

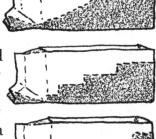

Let the soil containers dry out for about 15–30 minutes. After they have dried, prop the first carton against something immovable so that it is at a slight slant. Place a flat container under the spout opening so that it rests on the bottom of the pan.

Measure one cup of water and pour it steadily but gently on the top of the soil hill. Wait a few minutes for the water to settle and drain down into the pan.

Pour the water from the container back into the measuring cup and see how much you have gotten back, or recovered. Repeat this activity with the other two soil boxes, and again record your results. Also, make a note of just how clear or muddy the water is, or how much sediment or soil particles are present, as well as how long it took for the water you recovered to flow from each soil box into the drainage container.

What happens:

In our experiment, we recovered a cup of water from the step and plain boxes but half a cup from the box with gullies. There was a greater amount of sediment in the box without ridges than in the boxes with steps and gullies. Is that what you found?

Why:

Boxes that keep more of their soil in the boxes and not in the water are definitely experimental winners. Soil erosion, or the wearing away of topsoil, can be lessened by good farming or conservation techniques designed to protect and save the soil. The two such methods used in our experiment were contour farming, in which gullies or horizontal ridges are dug, or terracing, in which steps or elevated planes or levels are formed.

GRAVITY AND MAGNETISM: ATTRACTIVE FORCES

Although gravity and magnetism are different Earth forces, they both exert a lot of pull.

Gravity is the force that pulls everything downwards, towards the middle of our planet—you, your house, a ball, your bed, your car—everything! Your weight on Earth is simply the amount of pull this force has on you.

The planets, sun, and moon also have gravities, but with a force lesser or greater than Earth's. The sun's gravity holds the Earth and other planets in orbit around it, while the moon's pull lowers and raises the oceans' tides. Sir Isaac Newton, an English scientist, discovered these and other laws about gravity.

On the other hand, magnets have polar or field forces where an attraction, or pull, is stronger. The Earth itself, due to its iron center or core, is also a giant magnet.

The wonderful gravity and magnetic experiments here are bound to attract your attention and pull you away from anything else you're doing.

Rapid Transit

City subways or monorail train are often called rapid transit. Now watch how rapidly a ball will transit or move out of a tumbler, and learn about an important Earth force.

You need:
plastic tumbler
a small sphere (ball of
   clay, toy ball, marble)

What to do:
Place the sphere or ball in the tumbler and rapidly slide the glass, open end forward, across a tabletop or hard-surfaced floor. Stop the movement suddenly and observe what happens to the ball inside.

What happens:
The ball shoots out of the end of the stopped tumbler and keeps on rolling straight until something in its way stops it or changes its direction.

Why:
Sir Isaac Newton, an English physicist, discovered several natural laws of gravity and motion. One such law is called inertia. This means that something that is at rest will stay at rest, not moving, until another force works on it or moves it—and it will continue to stay in motion until, again, something works on it to stop it!

The ball in the moving tumbler stayed in it as long as it was moving. The inertia of this force was not overcome until the movement was suddenly stopped. The sudden stop was the force that overcame the inertia of the ball in the moving cup and sent the ball rolling—until a counter-force stopped it.

Weight Lifter

Weight is simply the pull of the force of gravity on you and on other objects. This experiment will demonstrate how this works. To avoid the mess of spills (gravity again), it's a good idea to do this activity outdoors.

You need:
waxed carton bottom
a thick rubber band
heavy string or twine
ruler
sharp pencil or nail
paper clip
a helper
bags or cups of substances to weigh
    (stones, gravel, beans, rice, dirt, sand,
    marshmallows)

What to do:
With the pencil or nail, poke a hole through one side of the carton about an inch (2 cm) down from the top and another hole directly across from the first, on the other side of the carton. Thread the ends of the string or twine through the holes and tie them securely to form a handle. Attach the paper clip to the top of the string handle and the rubber band to the other end of the clip.

Have your helper hold your homemade spring scale so that the top of the carton basket is even with the top of the ruler. From the substances available, select one and pour some of the gravel, stones, rice, dried beans, or whatever you wish to weigh into the carton. Do this slowly and gradually as you fill the carton.

Hold the top of the ruler to the top of the cup and calibrate, or measure, how many

inches or centimeters the gradually filled cup passes as it drops past the ruler.

What happens:

Your homemade spring scale, filled with different amounts of weight, measures the force of gravitational pull on the material in the basket. The carton basket is pulled down past the ruler's measurements according to the amount of force gravity exerts on it.

Why:

The Earth pulls everything towards its center. The more pull gravity is able to exert on an object, based on its denseness or mass, the heavier that object is. As the basket is filled, and the rubber band stretches, the amount of force measured by the spring scale grows.

Canned Laughter

Roll a can uphill and play a trick on a friend while learning about an important force that affects everybody and everything on Earth.

You need:

coffee can with lid
clay ball (golf ball size)
pencil

2 large (hardcovered, thick) books
an audience

What to do:

Place one end of a book on top of the other end of the other book to form a ramp. Place the clay ball inside the can and press it firmly against the side so it sticks to the can's surface. The ball should be "centered" on the can's wall somewhere between the two ends.

On the outside of the can, mark the spot with a pencil so that you'll know where the weight of the clay inside is concentrated.

Now, put the plastic lid back on the can and get ready to amaze your audience with an uphill roll.

Position the can on the lower end of the book ramp and experiment with it until you get it to roll up the book ramp. Now, find some friends who are interested in seeing your amazing "scientific magic."

What happens:
The can, surprisingly, rolls up the slightly uphill book ramp.

Why:
All objects are pulled towards the Earth's center by a constant, strong force called gravity. The "center of gravity" of any object seems to be the particular place on it where all the weight of the object is "centered." At this one point, the object will balance rather than fall.

The clay ball placed inside the coffee can was enough to reposition the can's natural center of gravity. The added off-center weight allowed gravity to pull the can forward and up the ramp.

ROLL PLAY

Test this same experiment but now use different surfaces. What happens if you place the can in the same "weighted position" but on the "downhill" side of the ramp? What happens if you place it on a flat surface?

Take the lid off the can and watch what happens to the clay weight inside it as you try different things. When the can does not roll, the weight in the can is concentrated in one place. When the center of gravity is shifted, the can is forced by the weight to move.

Where is the center of gravity? Press one end of a short string or piece of thread into the clay ball and watch.

The Dancing Cobra

This experimental trick done with a pin and a magnet will remind you of an Indian snake charmer and his swaying cobra.

You need:
cotton thread, about 8
  inches (20 cm) long
straight pin
bar or horseshoe magnet

What to do:
Make a loop in the thread and tie it around the head of the pin.

Hold the end of the thread with the pin attached and, with the other hand, lift it with the magnet. When you get the pin to an upright position, carefully lift the magnet from the pin so it is slightly suspended in midair. Move the magnet slowly in circles and watch the pin and thread, or "cobra," follow the movements. Unless you have a very strong magnet, there may only be a very small distance or break between the pin and the magnet or else the pin and thread will fall.

What happens:
The pin and thread floats suspended in the air slightly below the magnet and follows its path as you move it around.

Why:
The pin seems slightly to be overcoming gravity, floating below the magnet while not touching it. But this is proof that the magnet's attraction can pass through air and, at the right distance, can "balance" the force of gravity.

Don't Needle Me!

Make a magnetic compass that doesn't look like the usual one. It has no case and you won't have to needle it.

You need:
a very sharp pencil
U-shaped magnet
a large piece of
    modelling clay
    (to make a stand)
pencil

What to do:
Roll the piece of clay into a ball and flatten it to make a sturdy stand; then push the eraser end of the pencil into the clay stand. Carefully balance the U-shaped magnet on the pencil lead.

What happens:
The magnet gradually positions itself into a north–south direction.

Why:
The Earth is a magnetic ball with north and south magnetic poles. The U-shaped magnet positioned itself in a north-south direction because magnetic metals and liquids buried within the Earth's core have turned it into a giant magnet that naturally attracts all compasses and magnets. These great magnetic forces are concentrated at its north and south magnetic poles, which, incidentally, are not exactly the same as the north and south poles we normally speak of, but they are in the same area.

Needlework on the *Santa María*

Christopher Columbus and other early mariners, or sailors, probably used a wondrous device to help them travel the seas out of sight of land—a magnetized needle floating in a bowl of water.

Modern seafarers now have access to several devices to help them navigate the oceans, even a system of space satellites surrounding the Earth. But let's take a close look at that earlier version of the modern compass and see what a simple sewing needle can do, other than keep you in stitches.

You need:

sewing needle

bar magnet

bowl filled with water

small piece of wax paper

scissors

What to do:
Magnetize a sewing needle by rubbing one end of it fifty times with the north end of the bar magnet. Do the

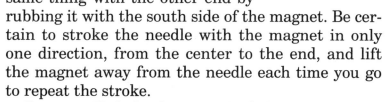

same thing with the other end by rubbing it with the south side of the magnet. Be certain to stroke the needle with the magnet in only one direction, from the center to the end, and lift the magnet away from the needle each time you go to repeat the stroke.

Cut a small circle about one inch (2 cm) in diameter out of the wax paper. Place the bowl of water on a table or kitchen counter. Stick the needle (be careful) into the wax-paper circle, as you would a needle into cloth. Float the wax paper with the needle on top on the surface in the middle of the water. Try to move it around on the surface. Observe what happens.

What happens:
The needle, when the movement stops, points north and south, no matter how many times you move it around.

Why:
Your floating compass needle is reacting to the Earth's invisible magnetic pull, caused by its giant bar-magnet core.

DON'T GET STUCK: CONTROL YOUR NEEDLE!

How do you know if all needles, when movement stops, position themselves in a north–south position? To find out, set up a control compass or one that lets you know if other things are causing results.

Do the same experiment in the same way but now substitute a non-magnetized needle for the magnetized one.

Move the compass to the middle of the water and again move the needle around. Wait patiently for the needle to stop moving. Do several trials or experiments and compare the control compass with the magnetized needle.

What's Your Point?

Magnetized straight pins with like poles repel each other, while unlike poles attract. True?

You need:

2 straight pins

2 paper clips

a magnet with north and
   south poles marked

cotton thread

a heavy book

What to do:

Magnetize a straight pin. Lay the pin on a hard surface and rub one end of it with the north side of the magnet. Rub from the center to the end, one way only, and lift the magnet between rubs. Do this about forty or fifty times. Repeat the rubbing action on the other end of the pin using the south end of the magnet. Magnetize the second pin the same way.

Write down which end (point or head) is north or south. Tie a thread to the center of each pin and attach paper clips to the other end. Dangle the two balancing pins about two inches (4 cm) apart, one on each thread, from the end of the table. Place the paper-clipped ends of the threads on the table and weight them down with the book. Now, try pushing the pins together.

What happens:

Some ends move away from each other, while the other ends jump at each other and bump.

Why:

Like magnetic poles repel or push away from each other while unlike poles attract or pull together.

Bartender

If you tend, pay attention, to this spinning bar magnet, you will notice some surprising things.

You need:
1 bar magnet
a long piece of cotton string
paper and pencil

What to do:
Tie one end of the long piece of string around the center of the bar magnet and tie the other end up someplace (light fixture, closet pole, rod between chair backs) where it can swing freely. Adjust the magnet so that it is properly balanced and does not hang down on one side.

Now, spin the magnet and wait about three or four minutes until the magnet has stopped moving. Draw a picture showing its north and south poles as they look to you. Do this five or six times. Will the magnet come to rest, or stop moving, in the exact same same position each time, with the poles aligned or the same?

What happens:
The bar magnet should continue to align itself up similarly, with the same poles showing, no matter how many times you spin it.

Why:
Hanging freely from a string, your bar magnet becomes a compass that aligns itself according to the Earth's magnetic pull.

Coin Artist

Many coin-operated food and soda machines, called vending machines, catch fake coins, such as slugs or washers, by using a magnetic part. How does this theft-preventer work?

You need:

4 real-money coins, such as U.S. dime, penny, quarter, nickel
2 or 3 metal washers

3 large hardcover books
1 bar magnet

What to do:

Stack two books on top of one another and lean the third against the stacked books to form a slide. Hold the bar magnet in the

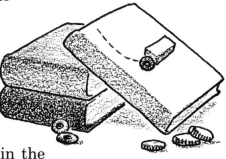

middle of the book forming the slide, while you drop each coin and washer down the side of the book, past the magnet.

What happens:
The coins slide past the magnet, but the washers are caught and held by it.

Why:
The bar magnet "picked up" the washers because they are made of steel or iron, but it did not pick up the official, government-minted coin, such as U.S. pennies and nickels, because these are minted, or made into coins, from alloys or mixed metals that are non-magnetic.

United States coinages are a combination of copper and other metals. Since magnets will not pick up copper, they are useful in catching all the slugs, or fake coins, made up of steel or iron that thieves may drop into vending machines.

Drawing Paper

Riddle: What kind of paper can you draw on, but yet never use a pencil nor be an artist? Do this experiment and find out.

You need:

steel wool pad
1 bar magnet
2 sheets of paper
an old scissors
a hand lens

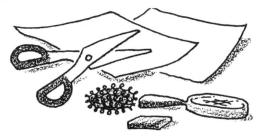

What to do:

On one sheet of paper, cut a steel wool pad into fine small threads. (Be careful of splinters.) Lay the bar magnet down and place the second sheet of paper over it so that the magnet is underneath in the middle. Now, carefully and evenly, pour the threads onto the sheet over the magnet. Lightly pound the table near the thread-covered sheet with your fist and watch the movement of the thread. Examine the thread patterns through the hand lens.

What happens:

The fine steel wool threads are drawn to and align themselves around the magnet in a circular pattern.

Why:

Definite circular lines of steel threads form around the magnet. This pattern is called the magnetic field of force. The steel threads gather more at the magnet's poles, where the force is greater, and thin out in the middle, where the force is less. This is the same magnetic force that encircles the Earth. Since the Earth is nothing more than a giant magnet, all steel and iron objects on its surface will behave this way.

DON'T FIDDLE WITH OLD FOSSILS

At one time or another, we've all left the lights on when leaving a room or turned on the air-conditioning when it wasn't *that* hot or the heater when it wasn't *that* cold. It's also very easy to forget to turn off the water promptly when finished, and even easier to just toss away aluminum cans, paper, glass and plastic bottles when they're empty—without thinking about it. But the Earth's resources, the natural substances we use to make energy and put it to work to make life better, are fast being used up. Some scientists predict Earth's resources will be gone within fifty years—your lifetime!

The Earth resources coal, oil, and petroleum are used to heat and cool our homes, make electricity, and fuel our cars. They are known as fossil fuels because they are made from the remains of ancient dead plants and animals. When these fuels are gone, they cannot be replaced.

There *are* other natural ways to create power, such as tapping into the heated water within the Earth (geothermal), nuclear energy, solar panels, windmills, or water-driven devices that move turbines (hydroelectricity) or geared machines that circle back to again produce more energy, etc. Until energy can be fully and inexpensively produced from such sources, however, fossil fuels will not be replaced.

So what can you do to help the Earth? Turn off lights when leaving a room, keep the thermostat or heater controls low so heat energy won't be wasted,

dress warmly or use a blanket when cold, drink cool water and wear light clothing when hot. Also, recycle paper, glass, aluminum, plastics, and metals (whatever is being collected where you are). Walk or ride your bike when you can, instead of taking a car. Don't waste water: Use less when washing, take shorter showers, don't flush unnecessarily, turn water off while you brush. Just for the record, one leaky faucet can waste thousands of gallons of water each year!

In this section you'll learn about conservation, fossil fuels, recycling, composting, and even learn to make your own recycled-paper note cards. But remember, it is your efforts to be a conservationist, an Earth-conscious, responsible person, that will help to save our planet, so don't fiddle with old fossils—save them!

HOUSEWARMING

A greenhouse is a closed glass house used to grow plants, where heat from the sun is trapped inside and moisture cannot escape.

Scientists see the Earth today as becoming a type of greenhouse. By burning coal, oil, and other such products known as fossil fuels, by over-using and abusing the use of our cars, and by heating and cooling our homes with electricity or gas, carbon dioxide and other harmful gases are being pushed into the atmosphere. These gases act as a dome, or lid, over the Earth's atmosphere, trapping the solar heat and preventing it from escaping into outer space. (See "Fossil Fuelish.")

When trees are cleared from large land areas, such as the tropical rain forests, tons more carbon dioxide gas remains in the atmosphere, instead of being converted into breathable oxygen. It's like putting the Earth into a big glass cooker, where heat from the sun is trapped and the air inside gets *hotter* and *stuffier.*

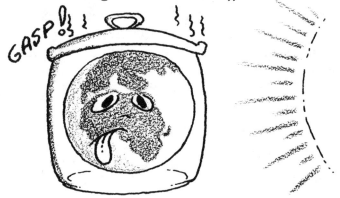

What Green House?

You may or may not find a greenhouse, or even a green house, on your block, but learning about the greenhouse effect and what it means to you and everything on Earth is very important today.

You need:
a glass container with cap or lid
1 teaspoon water
sunny outdoor location

What to do:
Place the teaspoon of water in the glass jar or bottle. Replace the lid or cap and tighten it well so that no air can escape. Leave the container outside in a sunny location for about an hour.

What happens:
Droplets of water form and cling to the sides of the container.

Why:
The sun's heat warms up the jar's atmosphere and the movement of the water molecules in it speeds

up. The water then evaporates into the air, but the moisture has nowhere to go, so it gathers into droplets, or condenses, on the cool glass sides. The lid on the jar acts as a greenhouse and produces the greenhouse effect. This is similar to the carbon dioxide gas that is produced by our own personal energy use and by the use of fossil fuels by industry which acts likes a lid over the Earth and prevents heat that is building up from escaping into space.

GREENHOUSE:
AN OPEN-AND-SHUT CASE

Do the same experiment as "What Green House?" but without putting on the cap or lid. Try the experiment with different-size containers (with lids and without) and with different amounts of water. Does the heat buildup that causes steam through evaporation occur earlier or later?

What differences do you notice? How do the experiments show what's happening to the Earth? Do the results suggest ways to prevent the greenhouse effect?

Warm Up

Measuring the heat energy trapped in a glass container shows us again how the greenhouse effect works and can affect us.

You need:

2 thermometers
2 strips of black construction paper or cloth, 2 by 6 inches long (5 x 15 cm)
a wide-mouth bottle or jar with lid
sunny outdoor location
paper and pencil

What to do:

Make sure that the temperature readings of both thermometers are the same—normal, outdoor temperature. Then find a sunny location outside and place the glass bottle on its side. Put a small rock or other weight against one or both sides of the bottle to keep it from rolling over.

Take one piece of black construction paper or cloth and place a thermometer on top of it, then slide both the thermometer and the material together into the bottle. Screw the lid on tightly but carefully so that the thermometer says in place.

Place the other thermometer on a dark strip next to the bottle. Record both temperatures, wait ten minutes, and record the temperatures again.

What happens:

The thermometer in the closed bottle registers a higher temperature than the one outside (in our experiment, there was a fifteen-degree difference).

Why:

This closed-environment experiment demonstrates the effect of large amounts of carbon dioxide (CO_2) in the Earth's atmosphere. CO_2 gas acts like the glass of the bottle, trapping heat. Although the sunlight falls equally on both sections of cloth, which absorb the light and produce the same amount of heat energy, the heat cannot easily radiate out through the glass barrier.

Carbon dioxide is produced naturally on Earth (given off as we breathe), but much more comes from the industrial burning of fossil fuels (smokestacks) and automobile engines, causing pollution and raising heat levels in the atmosphere. This "trapping" of heat by an increasing amount of CO_2 in the atmosphere is known as the greenhouse effect.

COOLDOWN

Try the experiment again, but leave out the black strips. Is there a difference? Bring your bottle with the thermometer inside it (don't open it) and the other thermometer indoors.

Write down the temperatures as the thermometers cool down. Is there a difference in the cooling-off time of the thermometer inside the bottle compared to the one outside? Which cools faster, and at what rate?

Oh, Ozone!

Make a model of the ozone layer, a thin layer of gas in the Earth's upper atmosphere that protects us from the sun's damaging ultraviolet rays.

Learn about CFCs, those chemicals that make life so much easier and better but yet do so much damage (they destroy ozone molecules). Then, watch as your ozone-layer model produces holes, gradually tears apart, and finally disappears!

You need:
1 short water or
    soda bottle
1 stick of chewing gum
very hot tap water
magnifying hand lens

What to do:
Chew the stick of gum
thoroughly. When it is soft, take it out of your mouth. Flatten it into a small disk between your fingers because you need a thin flat cap to seal the top of the bottle.

Now, fill the bottle *right to the top* with very hot tap water. Take the flat piece of gum and place it over the top of the bottle to seal it. Try to avoid making any holes, and make sure that the gum cap touches the water slightly. Observe what happens closely with the magnifying hand lens.

What happens:
The gum cap, as it touches the hot water, loses its elasticity, or stretchability, and holes begin to form. Eventually the gum cap breaks apart.

Why:

The bottle represents the Earth in our ozone model while the gum cap represents the ozone layer. The hot water touching the gum cap stands for the CFCs (chlorofluorocarbons), or chemicals that can damage ozone molecules.

CFCs are found in coolants for air conditioners and refrigerators and in the foam-plastic packaging used by some fast-food restaurants. These chemicals are released into the atmosphere as chlorine gas, which eventually destroys ozone.

So, cool it! But do cut down on air conditioner use, and do remind restaurant and business owners that they should act responsibly—that CFCs are harmful and that other types of food packaging are available.

STICK TO IT!

How can you reduce the amount of CFCs in Earth's atmosphere? Of course, you can't do it all alone, but you can do your part. Buy fewer products with CFCs in them, use less air conditioning, and remind others of our responsibilities to Mother Nature. Working together is the way to help save our Earth.

Now, do the same ozone experiment, but instead of filling the bottle to the top with hot water, stop when it is only half full, or less. Does the gum cap still show signs of wearing away? Is there a difference? Now you can see how releasing fewer CFCs into the air, or none, can make a big difference to our ozone layer.

RECYCLERS: HEAVYWEIGHT BOXERS

Recycling is the process of taking something and reusing it in the same or a different form. Plastic bottles and aluminum cans are melted down and made into new containers or other products while paper is returned to the pulp stage (a gooey, wet substance) and remade into new paper goods.

However, before you start tossing items into recycling bins or putting them out for collection, think of how you can reuse some of them yourself and cut down on "throwaway" products that can't be reused but only end up as pollution. Only buy products in packaging that is biodegradable (breaks down) or is recyclable. Look for packaging with the symbol of three folding arrows. This means the product was made from recycled goods or is collected and will be reused.

Give things such as toys, old clothes, appliances, furniture away to someone who can use them or to non-profit organizations that repair and distribute them to help others.

If you do not recycle paper, plastic, and aluminum yet (in many places, it is the law), you should start soon. Besides earning some pocket money, you and your family's efforts will help make the Earth a better, safer, and cleaner place to live.

My Recycling Center

Recycling is not hard—it's just a matter of getting organized and getting into the habit. So, get serious! Start now! Locate some strong, corrugated store cartons (these brown cardboard boxes are also recyclable when they have seen better days) and set up your own personal recycling bins.

You need:

a clean, dry location (basement, garage, or closed porch)

large, strong store boxes or wooden crates (ask at liquor stores or supermarkets)

marking pen for labelling

What to do:

Start by recycling easy things like aluminum and tin cans, glass bottles, newspapers, and plastic soda bottles. Some kinds of plastics are hard to recycle, such as soda bottles made with polyethylene terephthalate, or PET. Other plastics may be made up of several ingredients that cannot be separated for reuse and are not totally biodegradable. Scientists, however, are working on plastics that fully break down when exposed to sunlight (photodegradable) and bacteria to cut down on this pollution of our environment.

To be recycled, all materials should be free of contaminating food, dirt, or other substances. Label your boxes and sort materials. Remove all caps and lids from bottles and jars (they are made from different materials). Some places ask that brown and green glass containers be separate from clear glass, so you might keep them in separate bins.

Check to see if bottles are returnable for refund of

the deposit paid when purchased. Some areas have laws called "bottle bills" and you can earn some money by returning certain bottles and aluminum cans to stores that carry the product or to recycling centers.

Now that you've sorted out your materials and know what goes where in your mini-recycling center, it will be easy to get into the habit of recycling and you'll look forward to adding your efforts to the periodic recycling collections.

If you have a curb recycling program in your neighborhood, keep doing what is necessary according to the collection needs. If there are no regular pickups where you live, contact your nearest recycling center and ask about a scheduled pickup or when you can arrange to deliver what you've collected.

It's a Solar System

We're not talking about the sun, moon, and planets here, but rather about a solar water heater. In warmer parts of the world, a system of panels can be seen on the roofs of buildings and houses. These panels collect the sun's rays and use the energy to heat water.

Now it's time to make your own solar collector to heat water. It's fun, easy, and it won't take a lot of expensive equipment either.

You need:

8 feet (265 cm) aquarium air-line tubing (found in aquarium shops or store pet departments at low cost)
rubber band

a short wide-mouth jar or bottle
aluminum foil
an aluminum pan
a 2-litre soda bottle
sunny table outside

What to do:

Coil the plastic tubing back and forth accordion style leaving about a foot and a half (48 cm) of tubing loose at each end. Place the rubber band around the middle of the tubing, then shove the bunched tubing into the short container. Cover it with the foil, sealing the open jar around the tubing. Place the wrapped bottle in a pan on a table outdoors to preheat.

(This experiment should be done outside in a warm place where the rays of the sun will fall directly on your solar collector. The best time for the experiment is between 1 and 2 p.m., when the sun is at its strongest. Also, it is important to "preheat" the collector bottle by letting it sit in the sun 30 to 60 minutes before the experiment begins.)

To complete the experiment, fill the two-litre soda

bottle with cool tap water and place one end of the tubing into it. The bottle should be next to your solar collector on the table or bench. The other free tube should hang below the table.

Now, to get the water flowing through the collector bottle and out the free end, suck slightly on the end of the tubing as you would on a straw. This should start the water moving from the bottle through the collector and down the tube hanging below the table. The water will leave this tube in a slow, steady drip.

What happens:
The water that drips slowly from the tube towards the ground will be slightly but noticeably warmer than the water in the bottle.

Why:
Your solar heater is a miniature version of the large solar panels mounted on rooftops. Like the large collectors, your small model captures the sun's energy and heats the water travelling through the tubing. How much warmer the water will get, passing through your collector bottle, depends on a lot of things: the time of year, the time of day, the outside temperature, the location of the collector, how fast or slow the water passes through the system, and how long the collector is allowed to "preheat" before the water starts flowing.

Hot Under the Collar

What's hot under this bottle's collar, or opening? Now's the time to find out.

In "It's a Solar System" you really did two experiments. You built a solar collector, but you also made something called a siphon, a device used to draw off liquids from a higher to a lower place. A siphon works because of an important Earth force—gravity!

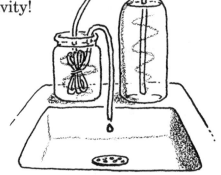

You need:
solar bottle collector
2-litre soda bottle
very hot and very cool
    tap water
use of sink and counter-
    top or table outdoors

What to do:
Remove the foil from the collector bottle. Leave the tubing as it is, and fill the bottle with very hot tap water. Fill the soda bottle with very cool tap water.

Place the two-litre soda bottle next to the collector bottle on the sink or small table outside. Insert one end of the free tubing into the bottle filled with cool water. The other end should hang down into the sink or towards the ground. Again, suck on the end of the lower tube to get your water moving.

What do you observe? Is the dripping water warmer than the water that came from your solar collector? What things, or variables, could have affected the temperature of the water coming through your collector?

Can you think of something you may have in your home that is similar to this device? How about a water heater?

FOSSIL FUELISH

Hundreds of millions of years ago, great plants like mosses and ferns growing in swamps died and fell one on top of the other. These layers became peat, or decayed plant matter. This process continued through the centuries until great beds, or layers, of decaying plants were formed and covered by mud, rock, and sediment. Pressed down by great weights, these layers were chemically changed by natural heat and pressure into beds of pure carbon, or coal, or became natural gas, mainly made up of a marsh or swamp gas called methane.

Crude oil was also formed in a similar way—chemically changed from the remains of small sea animals and plants by heat and great pressure. This compression, being tightly pressed together, turned the dead marine plant and animal matter into oil. Much of this chemical change took place under ancient seas—some of which no longer exist.

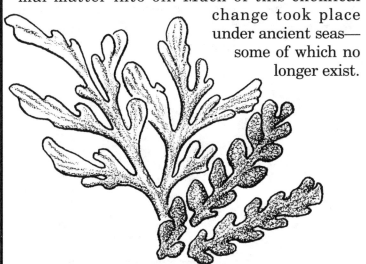

RAINWEAR

Is acid rain responsible for a gradual destruction of the world's forests? Perhaps not directly, but most scientists agree that any conditions that affect trees and plants are important to their health and our survival.

Acid rain is made up of nitric and sulfuric acids and produces poisonous metals, such as mercury, that can affect plants and animals. It contaminates the minerals in soil that plants need for food and the plants we need for food.

The sulfur and nitrogen elements of acid rain and acid dust, called acid deposition, also affect buildings, cars, statues, and other nonliving objects, causing much damage over the years.

These polluting acids can be found in rain, snow, fog, and the moisture available in the air in different strengths in all parts of the world. Did you know that some European cities have had rains that were as acid as lemon juice (2.3 on the pH scale)? How does the rain in your city compare? See "Spicy Test."

SPICY TEST

You can do an acid-base test to measure pH. In this test, liquids are applied to strips of test paper and the resultant color change is compared to a pH strip to determine just how acid or alkaline (base) a substance really is.

The pH scale is a range of numbers and colors. Number 1 is extremely acid while 14 is extremely alkaline, or base, which neutralizes acids. Seven is considered a neutral number.

Now, make your own test paper by using a spice called turmeric. Since the turmeric solution stains, cover your work area with newspaper; then, in a small cup, make a turmeric-paste solution. Add one tablespoon of turmeric to five tablespoons of hot tap water and stir until smooth.

Cut white construction paper (or thick paper towels) into small strips and dip them into the turmeric paste. Coat them well. The paste will stain your fingers but won't hurt you. Lay the golden, yellow-brown strips on newspaper to dry.

When thoroughly dried, test your paper in vinegar, soapy water, baking soda solution, lemon juice, or water with detergent. If the solution is very acidic, the strips will turn yellow; if alkaline, they'll turn a brownish-red. You can also use the turmeric test papers to test local tap water, lakes, rivers, soils, etc.

After testing, let your test papers dry and label them, telling what substances they were dipped into.

Papermaking: a Chip off the Old Block!

Paper is usually made of very tiny slivers of wood and water and processed to produce a pulpy mush. This is spread over a screen and left to dry. Now you can make your own paper simply and easily, and you won't even have to use a blender for chopping and mixing the glop or a regular deckle, a wooden frame with a wire cover over it.

This "recipe" makes four 6-inch (15 cm) square note cards.

Note: This activity is messy! Do the mixing on a kitchen counter and do the last part outside on a table that can be washed down. Also, be prepared! Paper making is a slow process. This activity may take all day to complete (e.g., soaking and drying).

You need:
aluminum foil
scissors
pencil
newspaper
large jar with lid
hot tap water
wooden spoon or spatula
aluminum pan
3 tablespoons cornstarch
½ cup hot water
markers, crayons, paints (for decoration)

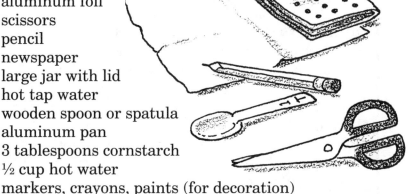

What to do:
Cut four pieces of aluminum foil and fold them over to form four 6-inch (15 cm) foil squares. They will be used to make a sieve, or simple deckle, to hold and drain the paper mixture. Punch holes with the

pencil in each of the foil squares, about a half-inch (1 cm) apart and in vertical rows.

Next, cut newspaper into long, thin strips and then cut or break them into smaller pieces. You'll need about 1½ packed cups of shredded paper.

Place the paper into the jar and fill it three-quarters full of hot tap water. Screw on the lid and let the mixture stand from two to three hours, shaking the jar from time to time and continuing to beat and stir the pieces with the wooden spoon to break them up. The more you stir and beat the pieces, the more the mixture will become pulpy, gloopy, creamier, and smoother. Add more water as the paper absorbs it.

When the mixture is pasty and creamy, pour it into the aluminum pan. Add a bit more water, if needed. Stir the mixture with the spoon again to be sure all the paper is broken up. Now, dissolve three tablespoons of corn starch into one half cup of hot water. Pour the solution into the paper mixture and mix thoroughly.

It's best to do the last part of this activity out-doors where the work area can be washed down. Place the tray with the mixture flat and put a foil square on top of it.

With the palms of your hands, press the foil downward until the mixture covers it. Bring the foil up and place it aside on the table. Press it flat with your hands to squeeze away the water. Repeat the process, using the other foil squares.

Lay a few sheets of newspaper in a sunny location and let your foil-backed paper dry.

As the paper dries, continue to press it down to squeeze out water. While you are doing this, pinch together any holes you notice.

After two to three hours, carefully peel the paper from its aluminum backing and trim it neatly into a square note card.

Now, pull out your crayons or paints, your colored markers, pencils,or chalk and design a special, homemade recycled card for Aunt Minnie, Grandpa Fred, or someone else you know. A specially made card from a fine, young, thoughtful conservationist like yourself is sure to be appreciated.

What now:

Your recycled newspaper looks and feels much like a gray egg carton. But, what would it look and feel like if we used different types of paper?

Collect scrap samples of used white writing paper and recycle it as you did the newspaper. What is the difference in the texture or feel and color of the paper?

116

EARTHY SUBJECTS

Composting with earthworms (vermicomposting) is a new and exciting way to recycle garbage. As a bonus, when earthworms eat dirt they produce valuable earthworm-enriched soil called castings, which is good fertilizer.

Maybe you can get a friend or family member to help out—someone who doesn't mind working with earthworms. If you get to know them, you may find other uses for these industrious creatures. You just never know how earthworms will measure up. They may even worm their way into the hearts of your entire family.

Earthworms—Starring in "A Real Dirty Deal!" (with a Cast of Thousands!)

If the idea of going to a bait shop and buying a pound of earthworms is the last thing you'd ever want to do, you are not alone. But before you say "No way," let's learn something about these helpful, earth-crawly creatures.

Earthworms have been grossly misunderstood, definitely getting the dirty end of the stick. Once you know something about them, however, you may begin to tolerate or even like worms. Then, we'll tell you how you can put them to work for you!

Earthworms lay eggs, have no eyes, and can regenerate, or grow, missing body parts. (Isn't that a really great trick!) Since they get the oxygen they need from the water and air available in the soil through their "skins," their bodies must stay moist, or wet.

Earthworms have short, stiff, hair-like legs, five hearts, and segmented, or separated, body sections. They take in, or eat, soil and pass it out as waste, a worm manure called castings.

This worm waste is great for fertilizing plants. It is full of nitrogen and minerals because worms eat fungi, bacteria, and old, decaying plant matter. This may not sound tempting to you, but plants like it a lot. Because of this, farmers and gardeners consider worm castings important and valuable, and they are often willing to pay a good price for it.

Now, maybe you won't mind working with worms, and putting them to work for you.

Worming Your Way into Composting

Vermicomposting ("vermi" means worm) is a great outdoor project. It can go on for as long as you like, and other family members can take part too. It's a very good way to recycle your garbage, and your garden plants will love the fertilizer your worms provide.

Wiggly worries? You don't have to be a scientist to work with worms, but you do have to be open-minded and adventurous. Composting with earthworms is not an exact science—there's really no right or wrong way of doing it, so you'll be learning along with the rest of us. The entire project is simply one big experiment to find out what works best for you. If something isn't quite right, try something different by adjusting the variables, things about the experiment that may or may not be making your project successful—and remember to keep notes.

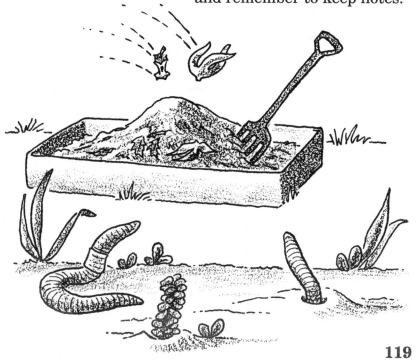

TEN TIPS FOR SUCCESSFUL VERMICOMPOSTING

To get you started, here are some helpful hints for keeping a worm colony odor-free and happy while producing fantastic plant fertilizer:

1 Select a cool place outdoors for your earthworm composting box (bin), and especially a place away from your house. This also will probably please others you live with, who might be less than excited about your unusual pets and new project. It is a good idea, too, to provide some sort of lid or covering for the box so birds and other animals won't be tempted to snack on your experiment-project workers.

2 Select a warm time of year to compost, such as late spring, summer, or early fall. Earthworms can't live in cold weather and prefer temperatures between 70 and 80 degrees F (about 21–26 C). One problem: In a regular compost bin or heap, bacteria break down or decompose materials and give off heat, making the bin hot. This is a natural part of regular composting, but the added heat is not good for earthworms. If the bin should get too hot, the worms will die and bacteria will become the main decomposers. Solution: Test the bin's temperature and, if it is hot, avoid adding the vegetable and fruit scraps, which could be encouraging the bacteria growth, for a few days to let the materials in the bin adjust.

3 Find a bin or box for your worms—not too deep (less than a foot) and about 2 feet (60 cm) square in size. A shallow box is especially preferred by Red Wigglers, a good choice for your bin, which are surface feeders. Bins can be made of plastic or wood, but must have holes in or near the bottom for ventilation and drainage. Wooden boxes, crates, and store-bought plastic garbage containers are perfect bins.

4 Bedding is needed in bins to keep earthworms moist, allow for air circulation, and for food. To your box, add 4 to 5 inches (10–13 cm) of shredded newspaper, followed by a 2-inch (5 cm) layer of soil, potting soil, manure, or a combination of these materials. Spray the bedding with water so it is damp, but not overly wet. An earthworm's body is made up mostly of water and its skin must be moist to take in oxygen and give off carbon dioxide. Too dry and it can't retain its life-giving moisture. But don't drown your earthworms. The bedding is wet enough when water starts flowing from the drainage holes. If overwatered, water can be squeezed from the bedding by hand.

Careful! Earthworms are harmed by chemicals that may be in water and paper. Avoid using colored (dyed) paper or water that is highly treated for bedding or extra moisture.

5 Red Wiggler worms are the best choice for composting bins. They can usually be bought at bait shops, or dug from yards with manure deposits and rotting leaf piles. They are surface feeders/breeders so will stay where they are supposed to. They are also hearty (won't sicken and die easily) and they reproduce, supplying you with new worms, quickly. When the bin is ready, you need only place them in the center of the box and wait for them to burrow down.

6 Generally speaking, a few pounds of worms (about 2,000) will eat about a pound of food a day, but don't worry if you only have ten, a hundred, or a thousand earthworms. Simply adjust the food to the number of worms and nature will do the rest. Everything will adjust in time, but it will probably take a while for the worms to decompose, or break down, the matter so that you can collect the castings. At this point, don't be afraid to experiment. Adjust food to worms and worms to food and patiently wait for results.

7 When feeding earthworms, chop or grind food scraps and bury them beneath the bedding to prevent gnats, flies, and the smell of rotting food (vermicomposting is supposed to be odor-free).

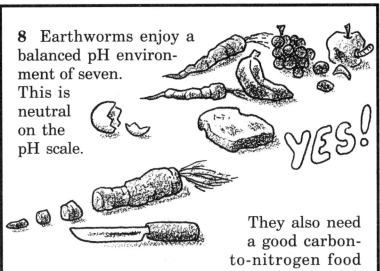

8 Earthworms enjoy a balanced pH environment of seven. This is neutral on the pH scale.

YES!

They also need a good carbon-to-nitrogen food mix. This means a carbon food such as shredded paper, and nitrogen food such as vegetable and fruit scraps will keep them happy.

If you still don't understand—not to worry! Simply feed the worms a balanced diet of grains, vegetable/fruit scraps and rinds, coffee grounds, and bread, while avoiding bones, yard vegetation, meat, fish, oily or acidy foods, and dairy products.

9 Since earthworms are pH neutral, you must avoid feeding them highly acidic foods such as tomatoes and citrus fruits. When acidic foods are used, balance them with crushed eggshells, which are alkaline and weaken acids.

10 Harvest or collect the castings or worm manure every month; too much of it in the bedding is not good for the worms. Push or pull the top layer of bedding to one side—most worms will be in the top third of the container material—and shovel the casting-rich bottom material from the box. Spread out the old bedding layer and add fresh bedding to the top as needed.

From time to time, gently turn over the bedding in your bin with a hand-held garden shovel or rake. You'll help circulate air among the colony and your earthy friends will love it.

INDEX